DNA TRANSFER TO CULTURED CELLS

Culture of Specialized Cells

Series Editor

R. Ian Freshney

DNA TRANSFER TO CULTURED CELLS

Edited by

Katya Ravid
Department of Biochemistry
Boston University Medical Center
Boston, Massachusetts

R. Ian Freshney
CRC Department of Medical Oncology
University of Glasgow
Glasgow, Scotland

Ⓦ**WILEY-LISS**
A JOHN WILEY & SONS, INC., PUBLICATION
New York • Chichester • Weinheim • Brisbane • Singapore • Toronto

Copyright © 1998 by Wiley-Liss. All rights reserved.

Published simultaneously in Canada.

While the authors, editors, and publishers believe that drug selection and dosage and the specification and usage of equipment and devices, as set forth in this book, are in accord with current recommendations and practice at the time of publication, they accept no legal responsibility for any errors or omissions, and make no warranty, express or implied, with respect to material contained herein. In view of ongoing research, equipment modifications, changes in governmental regulations and the constant flow of information relating to drug therapy, drug reactions, and the use of equipment and devices, the reader is urged to review and evaluate the information provided in the package insert or instructions for each drug, piece of equipment, or device for, among other things, any changes in the instructions or indication of dosage or usage and for added warnings and precautions.

Library of Congress Cataloging-in-Publication Data:

DNA transfer to cultured cells / edited by Katya Ravid, R. Ian
  Freshney.
      p.   cm.—(Culture of specialized cells)
    Includes index.
    "A Wiley-Liss publication."
    ISBN 0-471-16572-7 (pbk. : alk. paper)
    1. Genetic engineering—Methodology.   2. Transfection.   3. Cell
culture.   I. Ravid, Katya.   II. Freshney, R. Ian.   III. Series.
    [DNLM: 1. Transfection.   2. Cells, Cultured.   3. DNA—genetics.
QW 51 D629 1998]
QH442.6.D63   1998
571.9'648—dc21
DNLM/DLC
for Library of Congress   97-50456

Printed in the United States of America.

10 9 8 7 6 5 4 3 2 1

Contents

Preface

The ability to introduce novel genetic information into a cell and to detect expression of genes newly integrated in the genome was science fiction only decades ago. Since then, domestic, as well as foreign genes have been successfully introduced and expressed in a variety of eukaryotic cell types via a variety of constructs employing, among others, viral vectors and eukaryotic tissue-specific promoters. Several scientific advances have culminated in the ability to introduce, efficiently, one or a few copies of genes into different cell types. DNA transfer techniques are now widely used in many disciplines, including fundamental research in molecular genetics, regulation of development in normal and transformed cells, and in the biotechnology industry for the generation of biopharmaceuticals. One may also look forward to a time when this technology is even more widely available to those interested in the genetic modification of human cells for transplantation into genetically deficient hosts. Manipulation of a patient's own cells by restoration of one or more genes, provides an attractive alternative to allogeneic grafts and their attendant problems of epitope matching.

This is the fourth volume in the series "Culture of Specialized Cells" and has been prepared with the same objectives in mind, that is, to provide easily read practical protocols providing sufficient technical detail to complete the procedure without recourse to the prime literature. As in previous volumes, basic tissue culture techniques are not usually described in detail and these may be found in Freshney, [1994] "Culture of Animal Cells, a Manual of Basic Technique," Wiley-Liss, New York. Likewise, this volume contains many procedures which are basic to molecular biology, and a similar assumption is made that the reader will access such basic manuals as Ausubel et al., [1996] "Current Protocols in Molecular Biology," John Wiley & Sons, New York or Sambrook et al [1989] "Molecular Cloning, a Laboratory Manual," Cold Spring Harbor Press. It was not the intention to try to create another molecular biology manual but more to provide those working at the interface between tissue culture and molecular biology with the means to apply DNA transfer technology to cultured cells.

Different authors use different conventions for abbreviations and we have not attempted to standardize all of these. However, some have been standardized (e.g. UPW for ultra-pure water and PBSA for Ca^{+2} and Mg^{+2} free phosphate-buffered saline). These and other abbreviations are defined in a list following this Preface. The names of suppliers for specific reagents and materials are presented with each chapter, and the addresses, telephone, and fax numbers are provided at the end of the book. It is however, inevitable, that some of these names and numbers will become obsolete by the time the book is published, due to companies moving, merging, or going out of business. It is difficult to avoid this problem, as it can happen quite frequently, but updated information may be obtained from a number of web sites including <http://www.biosupplynet.com>, <http://www.informatik. uni-rostock.de/HUM-MOLGEN/biotech/companies/>, and <http:// www.ispex.ca/naccchembiochem.html>.

We are indebted to many of our colleagues for useful discussion and advice on the content of this book, and, of course, owe a great debt of gratitude to the contributing authors, who have provided excellent material to make up, what we hope, will be a very useful volume.

Katya Ravid
Ian Freshney

Contributors

Adams, Timothy E., Centre for Animal Biotechnology, School of Veterinary Science, The University of Melbourne, Parkville, Victoria 3052, Australia

Bichko, Vadim V., Scriptgen Pharmaceuticals, Inc., 200 Boston Avenue, Suite 3000 Medford, Massachusetts 02155, E-mail: bichko@virus1. fccc.edu

Brash, Douglas E., Department of Therapeutic Radiology, Yale School of Medicine, 15 York Street, New Haven, Connecticut 06520-8040

Cataldo, Leah M., Boston University School of Medicine, Department of Biochemistry, 80 East Concord Street, K225 Boston, Massachusetts 02118

Chen, Jin, Department of Microbiology and Immunology, Room AA4206 MCN, Vanderbilt University School of Medicine, 1161 21st Ave South, Nashville, Tennessee 37232-2363

Collodi, Paul, Department of Animal Sciences, Purdue University, West Lafayette, Indiana 47907

Conn, K. J., Department of Biochemistry, Boston University School of Medicine, 80 E. Concord St., Boston, Massachusetts 02118, E-mail: Foster@med-biochem.bu.edu.com

Craig, Ashley L., Department of Molecular and Cellular Pathology, Ninewells Medical School, University of Dundee, Dundee DD1 9SY, Scotland

Cuthbert, Andrew P., Human Cancer Genetics Unit, Department of Biology and Biochemistry, Brunel University, Uxbridge UB8 3PH, United Kingdom

Degterev, A., Department of Biochemistry, Boston University School of Medicine, 80 E. Concord St., Boston, Massachusetts 02118

Fontanilla, M. R., Department of Biochemistry, Boston University School of Medicine, 80 E. Concord St., Boston, Massachusetts 02118

Foster, J. A., Department of Biochemistry, Boston University School of Medicine, 80 E. Concord St., Boston, Massachusetts 02118, E-mail: Foster@med-biochem.bu.edu.com

Harris, Curtis C., Laboratory of Human Carcinogenesis, National Cancer Institute, National Institutes of Health, 37 Convent Drive, Building 37, Room 2C01, Bethesda, Maryland 20892-4255

Hicks, Geoffrey G., Manitoba Institute of Cell Biology and the Department of Physiology, University of Manitoba, 100 Olivia Street, Winnipeg, MB R3E OV9, Canada, E-mail: hicksgg@ctrvax.vanderbilt.edu

Hite, Jeffery P., Department of Molecular and Cellular Pathology, Ninewells Medical School, University of Dundee, Dundee DD1 9SY, Scotland

Hupp, Ted R., Department of Molecular and Cellular Pathology, Ninewells Medical School, University of Dundee, Dundee DD1 9SY, Scotland, E-mail: trhupp@bad.dundee.ac.uk

James, M. F., Department of Biochemistry, Boston University School of Medicine, 80 East Concord Street, Boston, Massachusetts 02118

Midgley, Carol A., Department of Biochemistry, Cancer Research Campaign Laboratories, University of Dundee, Dundee DD1 4HN, Scotland

Newbold, Robert F., Human Cancer Genetics Unit, Department of Biology and Biochemistry, Brunel University, Uxbridge UB8 3PH, United Kingdom, E-mail: margaret.kruger@brunel.ac.uk

O'Mahoney, John V., The Children's Medical Research Institute, Locked Bag 23, Wentworthville, NSW 2145, Australia

Quon, Michael J., Senior Investigator, Hypertension-Endocrine Branch, National Heart, Lung, and Blood Institute, National Institutes of Health, Building 10, Room 8C-103, 10 Center Drive, MSC 1754, Bethesda, Maryland 20892-1754, E-mail: quonm@gwgate.nhlbi.nih.gov

Ravid, Katya, Boston University School of Medicine, Department of Biochemistry, 80 East Concord Street, K225 Boston, Massachusetts

Reddel, Roger R., The Children's Medical Research Institute, 214 Hawkesbury Rd, Westmead, Sydney, NSW 2145 Australia, E-mail: rreddel@mail.usyd.edu.au

Rich, C. B., Department of Biochemistry, Boston University School of Medicine, 80 E. Concord St., Boston, Massachusetts 02118

Ricketts, Michael H., Department of Psychiatry, UMDNJ-Robert Wood Johnson Medical School, Piscataway, New Jersey 08854, E-mail: ricketts@umdrj.edu

Rosenberg, Robert, Department of Biology, Massachusetts Institute of Technology, 77 Massachusetts Avenue, Building 68-480, Cambridge, Massachusetts 02139-4307

Ruley, H. Earl, Department of Microbiology and Immunology, Room AA4206 MCN, Vanderbilt University School of Medicine, 1161 21st Ave South, Nashville, Tennessee 37232-2363

Schwartz, John J., Department of Biology, Massachusetts Institute of Technology, 77 Massachusetts Avenue, Building 68-480, Cambridge, Massachusetts 02139-4307, E-mail: jschwart@mit.edu

Strauss, William M., Department of Medicine, Division of Gerontology, Beth Israel Hospital, 330 Brookline Avenue, Boston, Massachusetts 02215

Trinkaus-Randall, V., Department of Biochemistry, Boston University School of Medicine, 80 East Concord Street, Boston, Massachusetts 02118

Wang, Zengyu, Boston University School of Medicine, Department of Biochemistry, 80 East Concord Street, K225 Boston, Massachusetts 02118

List of Abbreviations

δAg	Hepatitis delta antigen
δAg-L	Large form of hepatitis delta antigen
δAg-S	Small form of hepatitis delta antigen
μF	Microfarads
AcMNPV	*Autographa californica* Nuclear Polyhedrosis Virus
BES	*N,N*-Bis(2-hydroxyethyl)-2-aminoethanesulfonic acid
bGH	Bovine growth hormone gene
BRL	Buffalo rat liver cells
BSA	Bovine serum albumin
CAT	Chloramphenicol acetyltransferase
CHO	Chinese hamster ovary
CFU	Colony forming unit
CMV	Cytomegalovirus
CSLM	Confocal scanning laser microscopy
DAPI	4',6-Diamidino-2-phenylindole, a fluorescent marker for DNA
DDAB	Cationic lipid dimethyl dioctadecylammonium bromide
DDW	Deionized distilled water
DEAE	Diethylaminoethyl
DHFR	Dihydrofolate reductase
DMEM	Dulbecco's modification of Eagle's medium
DMEM-A	Dulbecco's modified Eagle's medium (BioFluids), pH 7.4 containing 25 mM glucose, 2 mM glutamine, 200 nM (R)-N^6-(1-methyl-2-phenylethyl)adenosine (PIA, Sigma), 100 μg/ml gentamicin, and 25 mM HEPES
DMEM-B	DMEM-A with BSA, 7% w/v BSA
DMF	Dimethylformamide
DMSO	Dimethylsulfoxide
DNA	Deoxyribonucleic acid
DOPE	Dioleoylphosphatidylethanol-amine
DOSPA	Polycationic lipid 2,3-dioleyloxy-*N*[2(sperminecarboxamido)ethyl]-*N,N*-dimethyl-1-propanaminium tryfluoroacetate
DOTMA	*N*-[1-(2,3 dioleyloxy)-propyl]-*n,n,n*-trimethylammonium-chloride.

DTT	Dithiothreitol
ECV	Extracellular virus particles
ED_{50}	Median effective dose
EDTA	Ethylenediaminetetraacetic acid
ES	Embryonic stem cell
F-12	Ham's F-12 medium
F12/FB	Ham's F12 media supplemented with 10% heat-inactivated fetal bovine serum and 5 μg/ml penicillin, 2.5 μg/ml streptomycin
FBS	Fetal bovine serum (=FCS in this context)
FCS	Fetal calf serum (=FBS in this context)
FDG	Fluorescein di-β-D-galactopyranoside
FGF	Fibroblast growth factor
FITC	Fluorescein isothiocyanate
G418	Geneticin
G418^R	Geneticin resistant
GFAP	Glial fibrillary acidic protein
GFP	Green fluorescent protein
gpt	Guanosine-phosphoribosyl transferase
GUS	β-Glucouronidase
Gy	Gray
HAT	Hypoxanthine aminopterin thymidine
HBS	HEPES-buffered saline buffer
HBSS	Hanks' balanced salt solution
HDF	Human dermal fibroblasts
HDV	Hepatitis δ virus
HEPES	4-(2-hydroxyethyl)-1-piperazine-ethanesulfonic acid
hGH	Human growth hormone
HMW	high molecular weight
IMDM	Iscove's Modified Dulbecco's Medium
IMDM/HS	IMDM with 2 mM glutamine and supplemented with 20% horse serum, and 5 μg/ml penicillin and 2.5 μg/ml streptomycin
kb	Kilobase
kbp	Kilobase pairs (of DNA)
KRBH	Krebs-Ringer medium, pH 7.4, containing 10 mM NaHCO$_3$, 30 mM HEPES, 200 nM adenosine, and 1% (w/v) bovine serum albumin
KRBH-A	KRBH with 5% bovine serum albumin
L-15	Leibowitz's L-15 medium
lacZ	β-galactosidase
LUC	Luciferase
mb	Megabase pairs

Mbp	Mega (million) base pairs
MMCT	Microcell-mediated monochromosome transfer
moi	Multiplicity of infection, or number of virus per cell
MoMLV,	Moloney murine leukemia virus
mRNA	Messenger RNA
NMBE	Normal human bronchial epithelial
NIH	National Institutes of Health
nRNP	Nuclear ribonucleoprotein complex
OV	Occluded virus particles
PBS	Phosphate buffered saline (Dulbecco and Vogt)
PBSA	Phosphate buffered saline, lacking Ca^{2+} and Mg^{2+}
PCR	Polymerase chain reaction
PDGF	Platelet-derived growth factor
PEG	Polyethylene glycol
PFG	Pulsed field gel electrophoresis
pfu	Plaque-forming units
PHA-P	Phytohemagglutinin
PIA	(R)-N^6-(1-Methyl-2-phenylethyl)adenosine
polh	Polyhedrin
RFLPs	Restriction fragment length polymorphisms
RNAse	Ribonuclease
RNP	Ribonucleoprotein complex
rpm	Revolutions per minute
RSV	Rous sarcoma virus
SCC-HN	Squamous cell carcinomas of the head and neck
SDS	Sodium dodecyl sulfate
SDS PAGE	Sodium dodecyl sulfate, denaturing polyacrylamide gel electrophoresis
Sf	*Spodoptera frugiperda*, the Fall army worm
SFM	Serum-free medium
SFM/CB	Serum-free DMEM with 5 μg/ml cytochalasin B
SFM/PEG	SFM containing 42.5% PEG (MW 1000) and 8.5% DMSO
SFM/PHA-P	SFM containing 100 μg/ml phytohemaglutinin
SHD	Syrian hamster dermal fibroblasts
SSC	Saline-sodium citrate buffer
β-gal	β-Galactosidase
STS	Sequence-tagged sites
SV40	Simian virus 40
TBE	Tris/borate EDTA buffer
TE	10 mM Tris HCl, 1 mM EDTA, pH 8.0
TEN (10x)	100 mM Tris HCl, 10 mM EDTA, 1 M NaCl, pH 8.0
TGF-β	Transforming growth factor-beta
TLC	Thin-layer chromatography

TNFM	4× SSC containing 0.05% Tween-20 and 5% nonfat milk
UPW	Ultra-pure water, eg, reverse osmosis or distillation combined with carbon filtration and high-grade deionization
UV	Ultraviolet light
VCM	Virus-conditioned medium
vRNP	Viral ribonucleoprotein complex
VSV	Vesicular stomatitis virus
X-Gal	4-Cl-5-Br-3-Indolyl-β-galactosidase
YAC	Yeast artificial chromosome

I

Production and Use of Retroviruses

Geoffrey G. Hicks, Jin Chen, and H. Earl Ruley

*Manitoba Institute of Cell Biology and the Department of Physiology,
University of Manitoba, 100 Olivia Street, Winnipeg, Canada (G.G.H.)
Department of Microbiology and Immunology, Vanderbilt University School
of Medicine, 1161 21st Ave South, Nashville, Tennessee 37232-2363
(J.C., H.E.R.)*

DNA Transfer to Cultured Cells, Edited by Katya Ravid and R. Ian Freshney.
ISBN 0-471-16572-7   © 1998 Wiley-Liss, Inc.

I. INTRODUCTION

Retroviruses are among the most competent vehicles available for introducing genes into mammalian cells. Efficiencies of gene transduction by retroviral vectors can approach 100%, and once integrated, the viral genome becomes a permanent genetic fixture in the DNA of the target cell. The genome of the host cell is largely unaltered and hence genetically identical to the parental cell. In addition, retrovirus infection occurs seamlessly as a sequence of normal cellular events and is in itself not harmful to the target cell. Together, these features make retrovirus transduction a superior vehicle for gene transfer as compared to other techniques, such as transfection, electroporation, microinjection, or lipid fusion.

For these same reasons our lab uses retroviral gene-trap vectors as insertional mutagens in embryonic stem cells (ES cell) [Chen et al., 1994; Hicks et al., 1995]. Disrupted genes are tagged by the inserted retrovirus and can therefore be cloned rapidly. Furthermore the mutagenized ES cells remain genetically competent for germline transmission.

More commonly, retrovirus vectors are used to transduce gene expression into mammalian cells where the biological consequences of the gene products can be examined. Comparative analysis with the otherwise genetically identical parental cell provides a powerful approach to study gene function. In this regard, retroviral vectors can provide gene function and have been used widely as vehicles for human gene therapy.

2. RETROVIRUS VECTORS

The general strategy for generating infectious retrovirus is outlined in Fig. 1.1. The process can be considered in four parts: first the construction of a retrovirus vector as a recombinant plasmid in *E. coli;* second, introduction of the plasmid into a packaging cell line; third, the incorporation of vector transcripts into transmissible virus particles; and fourth, conversion of the transcripts into double-stranded DNA by reverse transcriptase. At this time, vector sequences integrate into the genome of the infected cell. Protocols for producing stocks of transmissible recombinant retrovirus will be the main focus of this chapter.

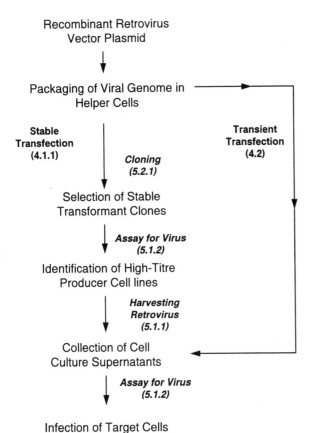

Recombinant Retrovirus
Vector Plasmid

Packaging of Viral Genome in
Helper Cells

Stable Transfection
(4.1.1)

Cloning
(5.2.1)

Transient Transfection
(4.2)

Selection of Stable
Transformant Clones

Assay for Virus
(5.1.2)

Identification of High-Titre
Producer Cell lines

Harvesting Retrovirus
(5.1.1)

Collection of Cell
Culture Supernatants

Assay for Virus
(5.1.2)

Infection of Target Cells

Fig. 1.1. Strategies for generating high-titer recombinant retroviruses. Overview of the approaches for packaging recombinant retrovirus constructs into infectious virus particles. Numbers in parentheses refer to the protocols detailed in this chapter.

Most recombinant retrovirus vectors are derived from the Moloney murine leukemia virus. A large portion of the viral genome encoding the gag, polymerase, and envelope genes are replaced with one or more mammalian gene expression cassettes (Fig. 1.2). The expression cassettes typically contain a suitable promoter, like SV40 or CMV; a selectable marker gene, like neo, hygro, puro, or lacZ; and a multiple cloning site into which cDNA for the gene of interest can be inserted. Two very successful and commonly used retroviral vectors are pBabeX [Morgenstern and Land, 1990] and pMFG [Dranoff et al., 1993]. These vectors are replication defective because they do not express the structural and enzymatic proteins required to produce infectious virus particles. This gives the user precise control over the process of gene delivery and expression, including multiplicity of infection, and the ability to maintain the normal biology of the target cell by preventing any pathology induced by a progressing retrovirus life cycle.

The vectors are packaged into infectious particles by expressing the recombinant viral genome in helper cells (Fig. 1.2). These pro-

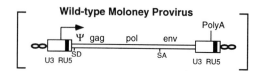

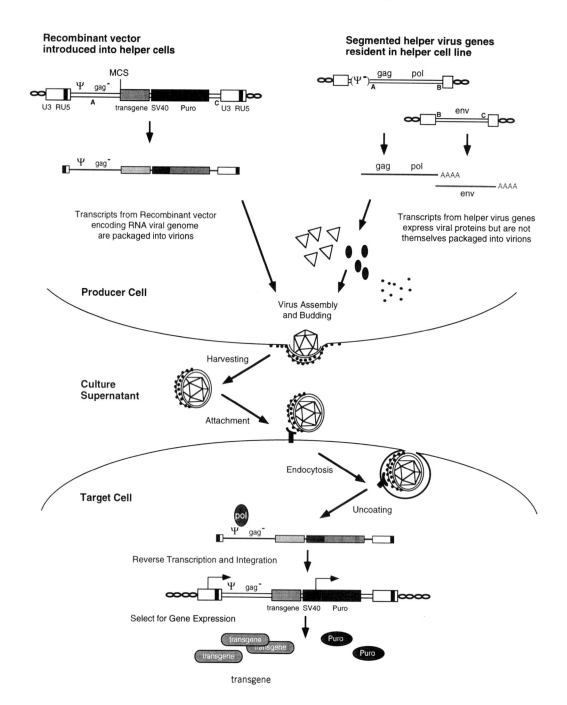

Hicks et al.

ducer cells have been engineered to express the appropriate viral packaging and structural proteins to complement defects in the recombinant vector. The genes expressing these viral proteins are encoded by separate recombinant plasmids integrated at different sites in the genome; this ensures helper viruses themselves are not packaged (Fig. 1.2). Infectious virions packaging the recombinant vector are assembled in the cytoplasm of the producer cell and bud into the surrounding culture medium from the cell surface. Here they can be harvested and stored as virus supernatants for later use.

Fig. 1.2. Retrovirus vectors and the helper functions contributed by packaging cells.

Packaging. Recombinant retrovirus vectors are introduced into helper cells. These vectors typically contain a mammalian gene-expression cassette replacing most of the body of the retrovirus. For example, pBabe-Puro (upper left construct) contains the puromycin resistance gene, *pac*, driven from the SV40 early promoter (SV40), which can identify target cells with stable integration of the recombinant virus. A multiple cloning site (MCS) permits the introduction of cDNA sequences for a given gene of interest (transgene). High-level gene expression in a broad spectrum of target cell types is achieved by the viral promoter in the long terminal repeat (LTR). Full-length transcripts (below) encode the RNA viral genome. A packaging signal (Ψ) and nearby group-specific antigen (gag) sequences acting in *cis* allow the transcripts to be packaged into virions with high efficiency. The gag protein itself is not translated because of mutations in the initiation codon (gag⁻). For comparison, the replication-competent Moloney leukemia virus is shown above. Long terminal repeats containing the viral promoter and enhancer functions are indicated, including the U3, R, and U5 regions (boxed). The initiation site of viral transcription (arrow), packaging signal (Ψ), splice donor (SD), splice acceptor (SA), and polyadenylation signals (PolyA) are also highlighted. In contrast to the recombinant vector, coding regions for the viral structural proteins, gag and envelope (env), and the viral reverse transcriptase/integrase (pol) remain intact. The proteins encoded by these genes are essential for the assembly of infectious viral particles. In the producer cell, these helper proteins are expressed from segmented viral genes resident in the packaging cell line. The choice of a producer cell line expressing a specific envelope protein (ecotropic, amphotropic, HIV, VSV-G, etc) will determine the host range of the virus. The formation of helper viruses is prevented by segmentation of the genes encoding virus proteins and extensive deletion of the virus' packaging signal (Ψ^-) and LTRs (open boxes). A minimum of three crossover events between the helper genes and the recombinant vector are required to regenerate a replication-competent helper virus. Sites of recombination are indicated (A, B, and C).

Harvesting. Once all the virus particle components are expressed in the packaging cell line (RNA genome, gag, pol, and env), infectious particles are assembled in the cytoplasm and bud from the cell surface into the surrounding culture media where they can be collected.

Infection. Retrovirus stocks are added to the culture media containing target cells for infection. For most retroviruses, cell attachment is mediated by specific binding to a receptor expressed on the cell surface. This interaction is enhanced *in vitro* with the addition of polyvalent cations to the media. Following endocytosis and uncoating of the virus particle, the viral RNA genome is reverse transcribed by the copackaged viral polymerase (pol). If the target cell is undergoing mitosis, a newly synthesized DNA template for the viral genome is integrated into the host cell DNA by the viral integrase (pol) and the retrovirus life cycle is completed. The target cell now constitutively expresses the transgene and the drug resistance gene, maintained by the viral and SV40 promoters, respectively. Collectively, the entire process described here results in the stable transduction of gene expression in a target cell.

Exploiting the biology of the murine retrovirus has resulted in technical advances that have increased the capabilities of Moloney-based vectors. For example, engineering the constitutive expression of the viral structural genes in the highly transfectable cell line, 293, allows for high-titer production of replication-defective recombinant viral vectors by transient transfection [Pear et al., 1993]. This protocol (3.2.1) overcomes several disadvantages of using retroviruses for gene transfer, namely, the production of high-titer retrovirus that lacks competent helper virus, vectors that can contain high transgene expression of a gene product which may be toxic to the producer cell, and most importantly, because of the protocol's rapidity, the generation of stable producer cell lines. Another major advance has been the ability to extend the range of host cells that can be infected. For example, the production of retroviral vectors with the VSV-G envelope protein (Section 4.3) permits the high efficiency required for gene transduction in human cells and the high titers required for retroviral use *in vivo* [Burns et al., 1993]. The only disadvantage of using retroviral vectors which has yet to be overcome is the restriction of their use to actively proliferating target cells. In this regard, adenoviral vectors maintain a distinct advantage.

With the biology of the retrovirus in mind, the packaging strategy should be designed with particular regard to the transduction efficiency or permissiveness of the target cell to retroviral infection. The primary limiting factor is the choice of envelope protein that will be sufficient for viral attachment and entry. For example, receptors for ecotropic envelope protein are widely expressed on mouse and rat cells; however, much greater transduction efficiencies can be obtained in hamster, human, or monkey cells by packaging virus with the amphotropic envelope. The use of VSV-G as a pseudotype envelope protein renders almost all eukaryotic cells permissive to infection, including fly, fish, frog, or human. These various approaches are described in the following section.

3. MEDIA AND STOCK REAGENTS

Dulbecco's modified Eagle's medium (DMEM), high glucose (10 liters).
   Powdered media for 10 liters

Sodium bicarbonate .... 18.5 g
Sodium pyruvate .... 1.1 g
Penicillin .... 2.0 g
Streptomycin .... 1.0 g

pH to 7.4 with HCl, sterile filter, and store at 4°C

gpt selection media in DMEM
Dialyzed FBS 10%
Xanthine .... 250 µg/ml
Mycophenolic acid .... 25 µg/ml
Aminopterin .... 2 µg/ml
Thymidine .... 6 µg/ml

Hanks' balanced salt solution, 0.1×
1× Hanks' balanced salt solution .... 1 part
Phosphate buffered saline (PBSA) .... 9 parts

Autoclave, and store at 4°C

TNE
Tris HCl, pH 7.8 .... 50 mM
NaCl .... 130 mM
EDTA .... 1 mM

Autoclave, and store at 4°C

Transfection buffer: 2× HBS
HEPES (pH 7.0) .... 50 mM
KCl .... 10 mM
Dextrose .... 12 mM
NaCl .... 280 mM
Na_2HPO_4 .... 1.5 mM

Filter-sterilize and store at room temperature.
All solutions in double distilled water (DDW).

4. PACKAGING

Recombinant retrovirus vectors are constructed as a provirus contained in cloned plasmid DNA. When introduced into cells, the RNA viral genome is transcribed, and the transcripts are packaged into infectious viral particles. As the *gag, pol,* and *env* genes are missing from recombinant vectors, the packaging functions of these proteins were originally provided by a helper virus in the producer cell. With the development of Ψ2 helper cells, a mutation in the packaging signal of the helper virus eliminated the problem of the helper virus itself being packaged [Mann et al., 1983]. In subsequent packaging cell lines, such as ΨCRE (ecotropic) [Danos and Mulligan, 1988], GP+E-86 (ecotropic) [Markowitz et al., 1988], or ΨCRIP (amphotropic) [Danos and Mulligan, 1988], and PA317 (amphotropic) [Miller and Buttimore, 1986] helper functions are expressed as different gene segments to reduce the potential of generating infectious helper virus by recombination. These packaging lines reliably produce high titers of helper-free replication-defective recombinant retrovirus (10^6 to 10^7 cfu/ml) when the recombinant vector is stably introduced into these cells [Miller et al., 1993].

Certain recombinant vectors may express gene products that are toxic or destructive to the producer cell itself, eg, in vectors designed to give high expression of a growth suppressor or oncogene gene product. In addition, some envelope proteins produced by the packaging cell line are highly fusogenic and result in syncytium formation and cell death. In these situations, high-titer retrovirus can be efficiently packaged when the toxic gene product is transiently expressed in modified 293 cells as the final step in packaging.

Protocols for the packaging of high-titer retrovirus by all three of these approaches are detailed in the following sections.

4.1. Generation of a Stable Producer Cell Line

A stable producer cell is generated by the constitutive expression of viral genome transcripts in a helper packaging cell line. This can be achieved either by infection of one producer cell with the retrovirus from another (as outlined in Section 5.1) or by transfecting recombinant plasmid vectors (Section 4.1.1). The former approach is preferred if one wants to generate a stable producer cell line from a transiently produced virus (Section 4.2), or to introduce the virus into a new packaging cell line in order to change its tropism.

By generating stable producer cell lines, one can screen multiple cell clones to identify lines that produce highest titer viruses, and to ensure a consistent level of transgene expression following infection of target cells. Once identified, the cryopreserved producer cell lines provide an indefinite source of the particular retroviral vector.

4.1.1. Stable transfection of packaging cells

While there are many techniques available for introducing plasmid DNA into mammalian cells, calcium phosphate-DNA coprecipitation remains the best option for stable integration. Although several protocols have been described, the BES method [Chen and Okayama, 1987] routinely yields higher transfection efficiencies. The distinguishing feature of this technique is the low pH of the BES buffer, which permits calcium phosphate-DNA precipitates to form gradually in the medium overnight, eliminating the need to regulate the initial mixing of HBS solutions. This allows for a better and consistent precipitate formation and high transfection frequencies (10–50%). Consequently, treatment with agents such as chloroquine, DMSO, or glycerol to enhance DNA uptake are also unnecessary, thus improving cell viability.

Protocol 4.1.1

Reagents and Materials

Transfection solutions
- CaCl$_2$: Prepare at 2.5 M and filter sterilize through a 0.45 μm nitrocellulose filter. Store at $-20°$C.
- BBS, 2× (BES buffered saline)

BES (N,N-bis(2-hydroxyethyl)-2-aminoethanesulfonic
acid, Sigma) .... 50 mM
NaCl .... 280 mM
Na$_2$HPO$_4$ .... 1.5 mM

The solution is adjusted to pH 6.95 with HCl, filter sterilized and stored at $-20°$C.
Note: The pH of this solution is critical to the high efficiency of this protocol.

Cell culture
- Plate exponentially growing helper cells 24 h prior to transfection at 5 × 10^5 cells per 100-mm dish in 10 ml Dulbecco's modified Eagle's medium (DMEM) supplemented with 10% fetal bovine serum (FBS) (not heat-inactivated) (DMEM10FB).

Protocol

(a) At room temperature, prepare a 0.5 ml DNA solution containing 20 μg of supercoiled plasmid DNA in 0.25 mM CaCl$_2$.

(b) Mix the DNA solution with 0.5 ml of 2× BBS, and incubate at room temperature for 20 min.

Note: Prior to drug selection, newly produced retrovirus can spread horizontally through the monolayer by infecting non-transfected helper cells. To ensure selection for transfected cells in which the number of plasmids integrated is high, the cells should be co-transfected (1 : 10) with a drug resistance marker not carried by the retrovirus being packaged. This will reduce the number of cell clones that must be screened in order to identify a high-titer producer cell.

(c) Add the calcium phosphate-DNA solution (1 ml) dropwise on top of the medium covering the cells and swirl gently.

(d) Incubate overnight (15–24 h) at 35°C under 3% CO$_2$ (if unavailable, a normal 37°C incubator with 5% CO$_2$ atmosphere will also produce satisfactory results).

(e) Following overnight incubation, remove the medium and wash the cell monolayer twice with PBSA.

(f) Incubate in normal growth medium for 24 h at 37°C and 5% CO_2.

(g) Subculture the cells, split at an appropriate ratio ($>1:10$), and reseed according to the cloning protocol of choice.

(h) Add DMEM containing the appropriate selection drug 24 h later.

(i) Change medium every 3 days. Individual colonies can be visualized without the aid of a microscope in 10–14 days. The colonies are cloned and established as cell lines as outlined in Section 5.2.

(j) High-titer producer cell lines are identified from among the panel of established cell lines by titrating for infectious retrovirus production, as detailed in Section 5.1.2.

4.2. Production of High-Titer Retroviruses by Transient Transfection

While the generation of stable producer cell lines has definite advantages, the process is time-consuming, and certain transgenes can be toxic to the producer cell. To circumvent these problems, a transient transfection strategy has been developed that uses the BOSC 23 packaging cell line [Pear et al., 1993]. The BOSC 23 cell line was derived from the adenovirus-transformed human embryonic kidney 293 cell line. These 293 cells are highly transfectable and the cells are resistant to several inducers of programmed cell death. BOSC 23 cells contain two plasmids, pCRIPenv- (expressing the amphotropic MuLV envelope glycoprotein) and pCRIPgag-2 (expressing the gag and pol genes). Since the genes encoding Moloney helper functions are on different plasmids, more than two crossover events are required to generate infectious helper virus. By using BOSC 23 packaging cells and optimizing the transfection procedure, high titer ($>10^7$ virus particles/ml) helper-free retroviruses can be produced.

It should be noted that this technology has recently been advanced by the inclusion of Epstein-Barr virus sequences into the plasmid backbone of retroviral vector templates [Kinsella and Nolan, 1996]. These next generation vectors permit the plasmid to be maintained episomally, resulting in a producer cell line which can stably produce high titers of infectious virus for more than a month.

Protocol 4.2

Reagents and Materials

• $CaCl_2$: 2.0 M, sterilize by filtration, and store at room temperature.

• HBS, 2× (see Section 3.0)

- Growth medium: DMEM10FB
- Chloroquine medium: DMEM10FB containing 25 μM chloroquine
- Gpt selection medium: See Section 3
- Cell culture: BOSC 23 cells are maintained in gpt selection medium and are resistant to G418, gpt, and hygromycin as a result of previous transfections. For short-term culture, cells can also be grown in DMEM10FB.
- Petri dishes, 60 mm

Protocol

(a) Seed 2×10^6 cells in 4 ml DMEM10FB (not gpt selection media) into each 60-mm dish approximately 18–24 h prior to transfection. It is important to grow cells as a single subconfluent monolayer and to avoid clumping.

(b) Just prior to transfection, change media to 4 ml chloroquine medium. Chloroquine neutralizes lysosomal activity and increases titers 2- to 3-fold. It is important that the cells be nearly confluent before transfection, otherwise significant cell death will occur.

(c) Prepare CaCl$_2$-DNA precipitate by first adding 6–10 μg of vector plasmid DNA to a solution containing 31 μl of 2M CaCl$_2$, 250 μl 2$\times$ HBS in a final volume of 500 μl, adjusted with distilled water.

(d) To this solution, add 500 μl 2$\times$ HBS (pH 7.0) dropwise and mix the components by gentle bubbling. This is best achieved in a 14-ml polypropylene tube (round bottom) by using a sterile-plugged 1 ml pipette attached to a pipette-aid. *The efficiency of transfection is directly related to the resulting size of the precipitate particles; the precipitate should be fine—cloudy to the eye but clearly visible under the microscope. Immediately add this solution to the cells.*

(e) After 7–11 h, replace the medium with DMEM10FB. To avoid toxicity, the chloroquine treatment should not be longer than 12 h.

(f) Harvest retrovirus at 48 h after transfection. *For maximum titers, the cells should be confluent by 24 h after transfection.*

(g) Filter the supernatant containing infectious retrovirus through a 0.45-μm nitrocellulose membrane to remove producer cells and cellular debris.

This virus preparation can be used directly for infection. Virus stocks can also be aliquotted, frozen by placing tubes directly into a $-80°$C freezer, and stored. Viral titers of the resulting cell supernatants should be 10^6 to 10^7 cfu/ml. The viral titer will drop 2- to 5-fold following each repeated freeze-thaw.

4.3. Production of VSV-G Pseudotype Retrovirus

A relatively new method allows for the generation of high-titer retrovirus in which the envelope protein has been replaced with the G protein of the vesicular stomatasis virus (VSV) [Burns et al., 1993]. These pseudotyped virions infect target cells by endocytosis followed by acid-induced fusion with phospholipid components of the endosome [Matlin et al., 1982]. Since this process is not dependent on the presence of a specific cell surface receptor, VSV-G pseudotyped vectors have a very broad host-cell range. As a consequence pseudotyped retroviral vectors are able to efficiently introduce foreign genes into cells derived from zebrafish, hamsters, or humans [Hopkins, 1993]. In addition, VSV-G pseudotyped vectors can be concentrated by ultracentrifugation obtaining titers of $>10^9$ cfu/ml with minimal loss of infectivity [Burns et al., 1993; Yee et al., 1994]. Because of their infectivity and stability, VSV-G pseudotyped retroviruses show great promise as new vehicles for human gene therapy.

4.3.1. Pseudotyping with VSV-G

Expression of the fusogenic VSV-G protein on the cell surface results in syncytium formation and cell death. To circumvent this difficulty, the VSV-G protein is transiently expressed following transfection of the gene into the 293GP-23 packaging cell line. The desired 293GP-23 packaging cell lines expressing transcripts for the retroviral vector of interest can be generated either stably (infection or transfection of the packaging cells followed by cloning) or transiently (transient co-transfection with the VSV-G expressing plasmid). It should be noted that an alternative producer cell line harboring an inducible VSV-G gene has recently been reported [Ory et al., 1996], which makes transient transfection of the gene unnecessary. In all strategies, the transient burst of VSV-G protein provides the last component needed to produce infectious retrovirus particles. Supernatants containing 10^6 to 10^7 cfu/ml can be routinely collected 48–84 h later.

Protocol 4.3.1

Materials and Reagents
Cell culture
- Seed 1 × 10⁶ 293GP-23 cells in DMEM10FB into 25-cm² vented flasks. Incubate for 2 days to allow the cells to attach firmly and to reach 80% confluence.

Lipofection
For each flask, prepare at room temperature in polystyrene tubes:

Solution A.
   for transient transfection into stable 293GP-23-producer clones
       5.0 μg pHCMV-G plasmid

or

for transient cotransfection into 293GP-23 packaging cells
   4.0 μg recombinant viral vector plasmid
   1.0 μg pHCMV-G plasmid
Dilute to a final volume of 260 μl DMEM (no serum and no *antibiotics*).

Solution B.
   Dilute 24 μl LipofectAMINE (GIBCO) to a final volume of 240 μl in DMEM (no serum and *no antibiotics*).

- Mix solutions A and B together by combining and inverting the tube several times.
- Following an incubation at room temperature for 45 min, bring the total volume up to 2.6 ml with DMEM containing 0.5% serum. It is important not to include antibiotics in the medium at any point during lipofection. LipofectAMINE reagents greatly enhance the cellular import of these drugs resulting in reduced cell viability.

Growth medium
DMEM10FB without antibiotics

Protocol
(a) Remove media from the 293GP-23 cells and gently wash cells twice with warmed PBS.
(b) Add DNA/lipofectin complexes dropwise to the center of each flask.
(c) Return the cells to the incubator for 5 h.
(d) Following incubation, add growth medium to a final volume of 8 ml.
(e) Collect 7 ml of viral supernatant 36 h, 60 h, and 84 h after lipofection, replacing with an equal volume of medium previously equilibrated in the CO_2 incubator.
(f) Filter the supernatant through a 0.45-μm nitrocellulose membrane to remove producer cells and cellular debris.

The virus preparation can be used directly for infection. Virus stocks can also be aliquotted, frozen by placing tubes directly into a $-80°C$ freezer, and stored.

Virus production will drop dramatically with excessive syncytium formation and detachment from the flask surface.

Note: It should also be restated that VSV-G pseudotyped virions are highly stable and infectious to humans. Appropriate containment precautions should be strictly adhered to (BCL-2 or higher, consult the biohazard safety regulations and guidelines for your institution).

4.3.2. Concentration of pseudotyped retrovirus

VSV-G pseudotyped virus can be concentrated 100-fold by ultracentrifugation with minimal loss in infectivity. Hence, titers of

$>10^9$ cfu/ml can be achieved. This protocol is adapted from Yee et al., [1994].

Protocol 4.3.2

Reagents and Materials

Aseptically prepared
- Virus supernatants
- Centrifuge tubes, ultraclear, (14 × 89 mm) for the SW41 rotor. Sterilize overnight by UV irradiation in a laminar flow hood. Alternatively, use autoclavable Polyallomar tubes.
- TNE or 0.1× Hanks' balanced salt solution
- Culture plate, 24-well

Nonsterile equipment
- Centrifuge
- SW41 rotor precooled to 4°C

Protocol

(a) Thaw virus supernatants (Protocol 4.3.1) quickly in a 37°C water bath and transfer to centrifuge tubes.

(b) Pellet the virus by ultracentrifugation in a precooled SW41 rotor at 50,000 g (25,000 rpm) at 4°C for 90 min.

(c) Decant the culture medium, in a laminar flow hood, and drain the tubes well. A clear pellet should be visible. The centrifuge tubes can be positioned upright in a 24-well culture plate.

(d) Resuspend the virus pellet in 0.5–1.0% of the original volume of supernatant in either TNE or 0.1× Hanks' balanced salt solution. This is best achieved by adding the medium over the pellets and storing the tubes overnight at 4°C, covered and without agitation. The pellet may then be dispersed by gentle pipetting without significant loss of infectious virus.

(e) Remove aggregated virions or cellular debris which fail to resuspend by low speed centrifugation (4000 rpm) in a microcentrifuge at 4°C for 5 min. This step is recommended if the virus stock is to be used *in vivo*.

(f) The virus may be further concentrated by a second round of ultracentrifugation. The resuspended virus preparations are pooled and pelleted as described above.

For each round of ultracentrifugation, a 70- to 100-fold increase in virus titers can be expected.

The virus preparation should be aliquotted based on the expected concentration factor, and frozen by placing sealed tubes directly into a −80°C freezer. The actual titers of the virus preparations, before and after concentration, can be determined as outlined in Section 5.2.

5. INFECTION AND SELECTION

Once a high-titer producer cell line has been established, the use of retrovirus for gene transduction or insertional mutagenesis in mammalian cells is quite simple: add the viral stock to the medium. The techniques involved in using recombinant retroviruses *in vitro* include harvesting retrovirus from producer cell lines, infection of target cells from these stocks of viral supernatants, titration of infectious virus as determined by drug-resistant colony forming units, and the selection and cloning of successfully transduced target cells. Many of the techniques described in the following protocols may also be applied to the derivation and screening of high-titer producer cell lines generated in the preceding section.

5.1. Retrovirus Collection and Infection of Target Cells

The biology of retrovirus infection involves attachment of the virion to the cell surface, receptor-mediated endocytosis, uncoating of the virion, and reverse transcription of the viral genome by the copackaged viral polymerase. The DNA template for the virus genome is made double-stranded by the host cell machinery, and it is then competent for integration into the host genome if the target cell is actively dividing. When completed, the transgene will be a permanent locus in the target cell genome with no undefined genetic changes or pathology to the target cell itself.

The frequency of gene transduction associated with a given producer cell line depends on the rate of virus production, the preparation of virus stocks, and the efficiency of target cell infection. Each process can be greatly influenced by a number of factors including temperature, cell density, attachment, and culture conditions. Together, these will be reflected in the apparent infectious titer. Optimizing these conditions to maximize the yield of infectious virus is the subject of Section 6.

The following methods are designed to provide a standard measure of virus transduction with regard to the rate of virus production, as titrated in a standard assay using NIH 3T3 target cells. While these methods are not optimized for total virus yield, they provide normalized titers with which different producer cell lines can be compared. This is helpful when determining the highest-titer producer cell from among many cell clones and to ensure the continued integrity of a producer cell line over time.

5.1.1. Harvesting of retrovirus from stable producer cells
Protocol 5.1.1

Reagents and Materials
- DMEM10FB
- DMEM, supplemented with 1% FBS serum, and previously equilibrated in the incubator.

Protocol
(a) Grow virus-producer cells at 37°C in DMEM10FB in a humidified, 5% CO_2 atmosphere. Culture conditions may vary for each helper cell line, particularly with the choice of calf or fetal bovine serum supplement.
(b) Seed 2×10^6 cells from each producer cell line into a 100-mm dish.
(c) Replace the medium, after 24 h, with 2 ml of incubator-equilibrated DMEM supplemented with 1% FBS serum.
(d) Collect infectious retrovirus in the fresh culture supernatant for 2 h, and then filter through a 0.45 μm nitrocellulose membrane to remove producer cells and cellular debris.
(e) The virus preparation can be used directly for infection.
(f) Virus stocks can also be frozen by placing tubes directly into a $-80°C$ freezer, and stored for use later. Titers will drop 2- to 5-fold with each freeze-thaw cycle, hence the supernatant stocks should be aliquotted prior to freezing.

If a number of newly derived producer cell clones are being examined to identify high-titer stable producer lines, expand the colonies to duplicate wells of a 6-well dish. As each cell clone approaches confluence, infectious virus may be collected and frozen from one well, while the cell line is cryopreserved from the duplicate well. When all clones have been processed, the virus supernatants may be thawed and titrated as described.

5.1.2. Assay for virus production on NIH 3T3 cells
Protocol 5.1.2

Reagents and Materials
- DMEM10CS: DMEM supplemented with 10% calf serum (CS)
- Polybrene, 8 μg/ml (hexadimethrine bromide)
- Virus: Make serial dilutions of the virus stocks into DMEM10CS and 8 μg/ml Polybrene. Polybrene is a polyvalent cation used to promote viral attachment and cell entry. Serial dilutions can be conveniently performed in 24- or 12-well dishes. Consider the following as a guide when deciding on a range of dilutions to assay:

(i) *Known titers or similar vector/producer systems.* Determine dilution necessary to score 2000 resistant colonies in your assay. For example, a U3Neo gene-trap retrovirus from a stable $\Psi 2$ packaging cell line will have an average titer of 6×10^6 cfu/ml in this standard assay. A 1:3000 dilution will contain 2000 cfu/ml. Make a small stock at this concentration by 1:10 serial dilutions. From this stock, make five 1:4 serial dilutions to be assayed. This will provide a linear range of resistant colonies from which an accurate titer can be determined.

(ii) *Unknown titer with similar target cell, or known titer with new target cell.* Make 1:10 serial dilutions of the viral stock up to 10^{-7}. For assay, choose a range of four dilutions that will produce from 1 to 10^4 resistant colonies, based on your best estimates. For example, an MFG vector [Dranoff et al., 1993] is expected to have a 10-fold higher titer than a gene-trap vector on the same NIH 3T3 cell; a gene-trap vector is expected to have a 10-fold lower titer if packaged in GP + E-86 cells rather than $\Psi 2$ cells; and the same gene-trap vector packaged in $\Psi 2$ cells will have a 10-fold lower titer in lymphocytes and 50-fold lower titer in ES cells, when compared with NIH 3T3 cells.

(iii) *Screening producer cell clones to identify those clones producing the highest virus titers.* Choose two 1:10 serial dilutions that, with the best of conditions, should yield 100 and 10 resistant colonies, respectively. The titers for individual cell clones will be relative to each other in this assay. *Note:* The rate of virus production varies widely among individual clones. However, 2–20% of clones have titers 10-fold higher than average ($>5 \times 10^5$ cfu/ml/10^6 producer cells/h). Therefore, at least one high-titer producer line can be established starting from 20 independent clones. These cell lines should be reassayed with the standard protocol to determine the accurate titer.

- Fixative: 10% formaldehyde in PBSA
- Stain: Crystal violet, 0.1% in equal parts PBSA and ethanol

Protocol
(a) Seed NIH 3T3 target cells used for titrating viruses 24 h prior to infection at a density of 5×10^5 per 100-mm dish and culture at 37°C DMEM10CS in a humidified, 5% CO_2 atmosphere.

(b) Place retrovirus stocks, fresh or quickly thawed at 37°C, on ice to increase virus stability.

(c) After removing media from the target cells, add 1 ml of diluted virus stock drop-wise to the center of each dish in the presence of 8 μg/ml Polybrene.

(d)   Incubate the cells at 37°C for 1 h with occasional rocking to allow for virus attachment.

(e)   Add 9 ml of complete media (DMEM10CS) and incubate the cells for 36–48 h prior to growth in the appropriate drug selection media. Retroviruses are routinely titrated by selecting for a constitutively expressed gene such as a neomycin resistance gene. This measures number of transducing viruses as determined by colony formation.

(f)   After 10–14 days, drug-resistant colonies are fixed for 10 min in 10% formaldehyde, stained in 0.1% crystal violet for 10 min, rinsed in tap water, and counted. Colonies should average 5 mm in size. Noticeably smaller colonies (<50 cells) are likely to be seedlings derived from larger ones and should not be scored. Viral titers are determined by the number of colonies on a plate times the dilution and divided by the volume of diluted viral stock added to the plate. The titer will be in colony-forming units per milliliter.

5.2.   Cloning Target and Producer Cell Lines

Following the infection of a target population of cells, it is often desirable to isolate clonal cell lines. This is typically done following selection for drug resistance. Individual colonies are transferred to a new culture dish and established in continuous culture as a permanent cell line. This protocol can also be used to clone individual producer cell lines following DNA transfection.

5.2.1.   Cloning of adherent cells by cloning cylinders
Protocol 5.2.1

Reagents and Materials
Aseptically prepared
- PBSA
- Trypsin/EDTA: 0.05% trypsin in HBSS (Hanks' balanced salt solution) containing 0.53 mM EDTA
- Complete medium: DMEM10CS
- Selection medium containing the appropriate selection agent, e.g., 400 μg/ml active G418
- Petri dishes, 150 mm
- 24-well culture plate
- Conical centrifuge tube, 15 ml
- Cloning rings, glass, in Pyrex petri dishes autoclaved with a thin film of silicone grease spread on the bottom surface
- Pipette tips, plugged

Nonsterile equipment
- Fine-tipped marker pen (Nikon manufactures a marking device that substitutes for one of the objectives on an inverted microscope)
- Micropipette

Protocol

(a) Wash the cell monolayer 36–48 h after transfection or infection with PBSA and remove the cells from the dish by overlaying 1 ml of trypsin/EDTA solution and incubating at 37°C until the cells detach (1–5 min).

(b) Wash the cells gently from the plate by repeated pipetting and transfer to a sterile 15-ml conical tube.

(c) Add 5 ml of complete medium to the cells to inactivate the trypsin, and then pellet the cells by centrifugation at 500 g (~1000 rpm) for 5 min in a table-top centrifuge.

(d) Resuspend the cells in complete medium and replate at clonal densities onto 150-mm dishes. Choose dilutions which will yield, optimally, 20–30 drug-resistant colonies per plate. Where the plating efficiency is unknown, a series of dilutions may be chosen (see Section 5.1.2.).

(e) Replace the growth medium, 24 h following replating, with selection medium.

(f) Replace selection medium every 3 days.

(g) Colonies about 5 mm in diameter should be easily observed 10–14 days later.

(h) Choose well-defined and isolated colonies for isolation and indicate on the plate by tracing the circumference of the colony on the bottom of the plate with a dark marking pen.

(i) Remove medium from the dish and wash the cells with PBSA, leaving 1 or 2 ml to prevent the colonies from drying while uncovered in the hood.

(j) Using sterile forceps, place cloning cylinders upright on silicone grease in Pyrex petri dish.

(k) Then carefully place a cloning cylinder directly around each colony using the marker-pen trace as a guide. Gently press the cylinder into position on the surface of the plate to allow the small amount of grease on the edge of the ring to form a water tight seal. Avoid excessive use of grease as it will cover the cells within the cylinder.

(l) Using a micropipette with sterile plugged tips or sterile transfer pipettes, add 60 μl of trypsin to each cylinder, which is now forming a well around and above the colony to be cloned.

(m) Incubate for 5 min.

(n) Gently pipette trypsin up and down to remove cells from the plastic and generate a single-cell suspension.

Production and Use of Retroviruses   19

(o)   Transfer the cells to an individual well of a 24-well culture plate containing selection medium. Note the cell line and clone number on the lid above the edge of the well.

(p)   Trypsinize individual wells as they become confluent, and serially passage to expand the cell line.

Once established, the new cell line should be cryopreserved, and then further characterized.

5.2.2.   Cloning by limiting dilution

Rather than cloning from isolated colonies some investigators prefer to clone cells by limiting dilution prior to selection. This protocol requires more work at the beginning, but has the advantage of very little maintenance afterwards. It is the method of choice for cells that are typically cultured in suspension.

Protocol 5.2.2

Reagents and Materials
- Trypsin/EDTA: 0.25% trypsin in ImM EDTA
- Multichannel pipettor and reagent reservoirs

Protocol
(a)   Trypsinize target cells, 36–48 h after transfection or infection, into a single-cell suspension and count.

(b)   Perform 1:10 serial dilutions into tubes containing 10 ml of growth media until the last tube contains <50 cells.

(c)   Using a multichannel pipettor and reagent reservoirs, replate each dilution (6 ml) into a 96-well culture dish in 60-μl aliquots.

(d)   Following 24 h of incubation under normal culture conditions, add a second 60-μl aliquot of growth media containing 2$\times$ drug concentration to each well, again using a multichannel pipettor. An equal volume (now 120 μl) of 2$\times$ selection medium may be added 5–7 days later, or as required.

Growth of cells in individual wells will be indicated by slight acidification of the medium and a resulting color change that can be easily visualized without the aid of a microscope. The dilution of cells in which less than 10% of the wells score positive for cell growth will have a probability of greater than 95% of being derived from a single cell and represent a clonal population. Choose this dilution to expand and characterize.

5.3.   Cryopreservation of Established Cell Lines

As with any characterized cell line, it is important to preserve early passage cells, whether or not the cell line is a producer, target, or genetically altered. This reduces the possibility of genetic drift within the population and thereby helps to ensure that the cell lines will be phenotypically stable. Cryopreservation is the best way to achieve this. Cell viability is maintained by using a slow and controlled rate of temperature drop during the freezing process. The use of a high serum concentration and DMSO helps to prevent the formation of ice crystals during the process and preserves the integrity of cellular membranes, preventing rupture.

Protocol 5.3.1

Reagents and Materials
- Complete medium: DMEM10FB or DMEM10CS as appropriate
- Freezing medium: Add 20 ml of serum (40% final) to 24 ml of growth medium. Then add 6 ml of DMSO (12% final) and mix thoroughly. Store at −20°C.
- Trypsin EDTA
- Vials for freezer storage (cryovials)
- Freezing box or controlled rate cooler
- Centrifuge tube, 15 ml

Protocol
(a) Trypsinize a subconfluent monolayer of cells, pipette up and down to form a single-cell suspension, and inactivate the trypsin by gently mixing the cell suspension with a 5-fold excess of complete medium. Remove a small aliquot to determine cell number, and centrifuge the remaining cells at low speed to remove medium.

(b) Resuspend the cell pellet in freezing medium at room temperature to a concentration of 2×10^6 per ml, and aliquot into cryovials.

(c) Cool cryovials to −80°C at a rate of 1°C/min. This is accomplished by either the use of special freezing canisters (Nalgene) or by wrapping the tubes in cotton wool and then in small Styrofoam containers. Place the containers into a −80°C freezer and allow to cool for a minimum of 2 h. Transfer vials to regular freezer boxes or canes and store at −80°C or in liquid nitrogen vapor (gas phase of liquid nitrogen freezer).

(d) To return cryopreserved cells to culture, bring vials rapidly to 37°C by immersion in a water bath.

(e) Resuspend cells gently and transfer to a 15-ml conical tube and add an equal volume of prewarmed complete growth medium. To reduce loss of viability by hypotonic shock, increase the volume of media incrementally until a nearly normal serum concentration is achieved.

(f) DMSO induces differentiation in some cell types and should be removed by pelleting the cells in a low-speed centrifuge.

(g) Replate cells at a relatively high density.

(h) After 24 to 36 h, continue to culture cells under normal conditions for the cell line.

(i) Return to selection media if required.

6. METHODS TO INCREASE VIRAL TITERS

The protocols outlined in the previous sections reflect experimental conditions that provide a reproducible measure of virus titers. Such standardized methods are critical to the use of viral vectors as insertional mutagens and in the comparative analysis of newly developed vectors. These factors may not be of equal importance if retroviral vectors are to be used for the transduction of gene expression in mammalian cells, where often it is more desirable to obtain viral supernatant with the highest possible infectious titers.

In order to scale up virus production or to enhance relative viral titers in an already established producer cell line, it is important to consider a number of factors that affect infectious titers. While this section details factors which together might increase absolute titers by up to 100-fold [Lee et al., 1996], it should not replace careful deliberation concerning the initial choice of vector (best promoter for transgene expression) or envelope protein (permissiveness/efficiency of target cell to infection).

6.1. Viral Stability and Temperature

While retrovirus particles are stable at 4°C, they are surprisingly unstable at 37°C. To maintain the titer of viral supernatants once they have been collected it is important to keep stocks and dilutions on ice until they are either used to infect target cells (within 2 to 3 days) or frozen.

Freezing virus supernatants by placing tubes directly into a −80°C freezer is the most common method of preserving virus stocks. Each freeze-thaw cycle of a stock solution results in a 20–50% drop in infectious titers. There is no cryopreservative agent that appears to stabilize virions during this process. It is important therefore to store virus stocks in aliquots for later use, thereby avoiding repeated thawing. If virus titers are important to the subsequent experiments, the titer from a sample vial should be determined after thawing.

Producer cells maintained at 32°C may yield 50–100% more virus than cells maintained at 37°C due to the same temperature-dependent stabilization of the virions [Kotani et al., 1994; Lee et al., 1996]. However, one must also consider the growth of the producer cells at the lower temperatures. In this regard, incubation at 35°C may be the preferred compromise.

6.2. Production and Collection of Virus

Several factors may influence the production of infectious virus during harvesting. Generally, the highest production of virus from a given producer cell line will be from the point the monolayer reaches confluence and up to 5 days following. While we typically harvest virus 2 h after a medium change, it is possible to increase the relative titer of the supernatant between 25 and 50%, by harvesting at 15 h or overnight. The production of infectious virus remains relatively constant for up to 5 days after the producer cells have reached confluence. Hence, preparations of viral supernatant can be collected every 24 h without a significant drop in the average titer by replacing the conditioned media with fresh incubator-equilibrated media. For large-scale preparations of virus stocks, the use of roller bottles or perfusion culture permits greater volumes of high-titer supernatants to be collected.

When lengthening the harvesting times it becomes important to pay particular attention to the culture conditions. While one can increase relative titers by collecting the virus in a smaller volume of media, be sure to account for any added length of time in regard to optimum cell viability. If changing or perfusing the medium, use medium that has been equilibrated in the same incubator environment. This will ensure cell attachment to the culture dish matrix and maintain virus production.

Finally, while serum concentrations will not make a significant difference in the relative production of virus or viral stability, it may be significant in maintaining the optimum producer cell viability over extended harvesting times. Serum concentrations in the viral supernatant should also be considered if the supernatant will be further processed following collection (e.g., by concentration techniques using ultrafiltration). As an important final note, be sure to use the right kind of serum for the producer cell line being employed. Virus production in Ψ2 cells may be significantly reduced by using fetal bovine serum instead of calf serum.

6.3. Concentration of Viral Supernatants

Ultracentrifuging VSV-G pseudotyped virus can offer the tremendous advantage of increasing relative titers by over 2000-fold (Section 5.3.2). There are similar approaches for concentrating eco- and amphotropic viral vectors, such as polyethylene glycol precipitation or stirred ultrafiltration, which can increase relative titers 10- to 30-fold. Protocols for these methods have been described elsewhere [Ausubel et al., 1993].

6.4. Target Cells and Viral Infection

The most important factor to ensure high transduction efficiencies is to infect target cells that are rapidly proliferating. Because it is during mitosis that the retrovirus gains access to the nucleus [Lewis and Emerman, 1994; Miller et al., 1990], it is critical to infect exponentially growing cells.

High transduction efficiencies are also enhanced (100- to 1000-fold) by the use of polycations that promote viral attachment and cell entry. The use of Polybrene at 8 μg/ml typically gives titers that are several times better than titers obtained with alternative polycations such as protamine sulfate or poly-L-lysine. Although all polycations should be assayed for cellular toxicity, Polybrene is usually less toxic to a broader range of target cell types. For some cell types, the use of DEAE-dextran can increase relative titers 2- to 3-fold over Polybrene [Lee et al., 1996].

Finally, increasing the infection time of target cells to 24 h with viral supernatants may increase the relative titers 2- to 10-fold [Morgan et al., 1995]. Cell viability during prolonged culture in small volumes should be taken into consideration, as outlined in the preceding section.

ACKNOWLEDGMENTS

The authors would like to thank Chris Aiken for critical review of this manuscript, and members of Dr. Earl Ruley's lab, past and present, who have contributed directly or through the experience of others.

REFERENCES

Ausubel FM, Brent R, Kingston RE, Moore DD, Seidman JG, Smith JA, Struhl K (eds) (1993): Current Protocols in Molecular Biology. New York: John Wiley & Sons.

Burns JC, Friedmann T, Driever W, Burrascano M, Yee JK (1993): Vesicular stomatitis virus G glycoprotein pseudotyped retroviral vectors: Concentration to very high titer and efficient gene transfer into mammalian and non-mammalian cells. Proc Natl Acad Sci USA 90: 8033–8037.

Chen C, Okayama H (1987): High-efficiency transformation of mammalian cells by plasmid DNA. Mol Cell Biol 7: 2745–2752.

Chen J, Degregori J, Hicks GG, Roshon M, Shi E-G, Scherer C, Ruley HE (1994): Retrovirus gene traps. Methods Mol Genet 4: 123–140.

Danos O, Mulligan RC (1988): Safe and efficient generation of recombinant retroviruses with amphotropic and ecotropic host ranges. Proc Natl Acad Sci USA 85: 6460–6464.

Dranoff G, Jaffee E, Lazenby A, Golumbek P, Levitsky H, Brose K, Jackson V, Hamada H, Pardoll D, Mulligan RC (1993): Vaccination with irradiated tumor cells engineered to secrete murine granulocyte-macrophage colony-stimulating factor stimulates potent, specific, and long-lasting anti-tumor immunity. Proc Natl Acad Sci USA 90: 3539–3543.

Hicks GG, Shi E-G, Chen J, Roshon M, Williamson D, Scherer C, Ruley HE (1995): Retrovirus gene traps. Methods Enzymol 254: 263–275.

Hopkins N (1993): High titers of retrovirus (vesicular stomatitis virus) pseudo-types, at last. Proc Natl Acad Sci USA 90: 8759–8760.

Kinsella TM, Nolan GP (1996) Episomal vectors rapidly and stably produce high-titer recombinant retrovirus. Hum Gene Ther 7: 1405–1413.

Kotani H, Newton PBr, Zhang S, Chiang YL, Otto E, Weaver L, Blaese RM, Anderson WF, McGarrity GJ (1994): Improved methods of retroviral vector transduction and production for gene therapy. Hum Gene Ther 5: 19–28.

Lee SG, Kim S, Robbins PD, Kim BG (1996): Optimization of environmental factors for the production and handling of recombinant retrovirus. Appl Microbiol Biotechnol 45: 477–483.

Lewis PF, Emerman M (1994): Passage through mitosis is required for on-coretroviruses but not for the human immunodeficiency virus. J Virol 68: 510–516.

Mann R, Mulligan RC, Baltimore D (1983): Construction of a retrovirus packaging mutant and its use to produce helper-free defective retrovirus. Cell 33: 153–159.

Markowitz D, Goff S, Bank A (1988): A safe packaging line for gene transfer: separating viral genes on two different plasmids. J Virol 62: 1120–1124.

Matlin KS, Reggio H, Helenius A, Simons K (1982): Pathway of vesicular stomatitis virus entry leading to infection. J Mol Biol 156: 609–631.

Miller AD, Buttimore C (1986): Redesign of retrovirus packaging cell lines to avoid recombination leading to helper virus production. Mol Cell Biol 6: 2895–2902.

Miller AD, Miller DG, Garcia JV, Lynch CM (1993): Use of retroviral vectors for gene transfer and expression. Methods Enzymol 217: 581–599.

Miller DG, Adam MA, Miller AD (1990): Gene transfer by retrovirus vectors occurs only in cells that are actively replicating at the time of infection. Mol Cell Biol 10: 4239–4242.

Morgan JR, LeDoux JM, Snow RG, Tompkins RG, Yarmush ML (1995): Retrovirus infection: effect of time and target cell number. J Virol 69: 6994–7000.

Morgenstern JP, Land H (1990): Advanced mammalian gene transfer: high titer retroviral vectors with multiple drug selection markers and a complementary helper-free packaging cell line. Nucleic Acids Res 18: 3587–3596.

Ory DS, Neugeboren BA, Mulligan RC (1996): A stable human-derived packaging cell line for production of high titer retrovirus/vesicular stomatitis virus G pseudotypes. Proc Natl Acad Sci USA 93: 11400–11406.

Pear WS, Nolan GP, Scott ML, Baltimore D (1993): Production of high-titer helper-free retroviruses by transient transfection. Proc Natl Acad Sci USA 90: 8392–8396.

Yee JK, Friedmann T, Burns JC (1994): Generation of high-titer pseudo-typed retroviral vectors with very broad host range. Methods Cell Biol 43: 99–112.

APPENDIX: MATERIALS AND SUPPLIERS

Material	Supplier	Catalogue Number
Equipment		
Cloning cylinders	Belco	2090-00808
Cryo-freezing containers	Nalgene	5100-0001
Cryovials	Nunc	3-77267
Filters, Acrodisc, 0.45 μm	Gelman/VWR	4184/28144-007
Multichannel pipettor	Corning-Costar	
Petri dishes, 150 mm	Becton Dickinson	
Plates, 24-well	Becton Dickinson	
Reagent reservoirs	Corning-Costar	4870
Ring marking device	Nikon	
SW-41 rotor	Beckman	
Tubes, 14 × 89 mm	Becton-Dickinson	
Chemicals		
Aminopterin	Sigma	A-3411
BES	Sigma	B6137
Calf serum (donor)	Sigma	C-5280
Chloroquine	Sigma	C-6628
Dextrose (D-(+)-glucose)	Sigma	G7528
DMSO	Fisher	BP-2311
Dulbecco's modified Eagle medium	Sigma	D-5648
Fetal bovine serum	Sigma	F-4135
G418 (Geneticin)	Sigma	G-5013
Hanks' balanced salt solution	GIBCO	24020-042
HAT	Sigma	H-0262
HEPES	Sigma	H-7523
LipofectAMINE	GIBCO	18324-012
Mycophenolic acid	GIBCO	860-1814H
Penicillin	GIBCO	860-1830MJ
Phosphate buffered saline (PBSA)	GIBCO	310-4040AJ
Polybrene (hexadimethrine bromide)	Sigma	H-9268
Sodium bicarbonate	Sigma	S-5761
Sodium pyruvate	GIBCO	11840-030
Streptomycin sulfate	GIBCO	860-1860IC
Thymidine	Sigma	T-1895
Trypsin-EDTA solution	GIBCO	25300-054
Xanthine	Sigma	X-2001
Cells and Plasmids		
293GP-23 Cells	Chiron-Viagene	
Plasmid pHCMV-G	Gift from Jane Burns	

2

Baculovirus Expression and the Study of the Regulation of the Tumor Suppressor Protein p53

Carol A. Midgley, Ashley L. Craig, Jeffery P. Hite, and Ted R. Hupp

Department of Biochemistry, Cancer Research Campaign Laboratories, University of Dundee, Dundee DD1 4HN, Scotland, UK (C.A.M.). Department of Molecular and Cellular Pathology, Ninewells Medical School, University of Dundee, Dundee DD1 9SY, Scotland, UK (A.L.C., J.P.H., T.R.H.).

DNA Transfer to Cultured Cells, Edited by Katya Ravid and R. Ian Freshney.
ISBN 0-471-16572-7   © 1998 Wiley-Liss, Inc.

1. INTRODUCTION

In this chapter we describe the basic techniques for inserting the gene encoding a protein of interest into the baculovirus genome, in order to achieve high-level recombinant protein expression in insect cells. To illustrate the versatility of this expression system as a research tool we also summarize our work on the post-translational modification and regulation of p53 tumor suppressor protein expressed in insect cells.

Baculovirus expression vector technology is now a popular route to the production of high levels of biologically active eukaryotic proteins [Lucknow and Summers, 1988; King and Possee, 1992]. Several hundred genes have now been expressed with production efficiencies of between 1 and 500 mg/l. The nuclear polyhedrosis virus AcMNPV isolated from the alfalfa looper *Autographa californica* is the prototype of the family *Baculoviridae*. AcMNPV has been extensively studied because of its efficient replication in cell culture and is now the basis of most baculovirus expression vectors. Propagation of baculovirus was originally performed in insect larvae, but as it is relatively easy to establish cultures of insect cells *in vitro,* the majority of procedures are now carried out in cell lines derived from the Fall army worm *Spodoptera frugiperda* (Sf). In the gut of insect larvae and in cultured cells, baculoviruses reproduce in two forms (Fig. 2.1); extracellular virus

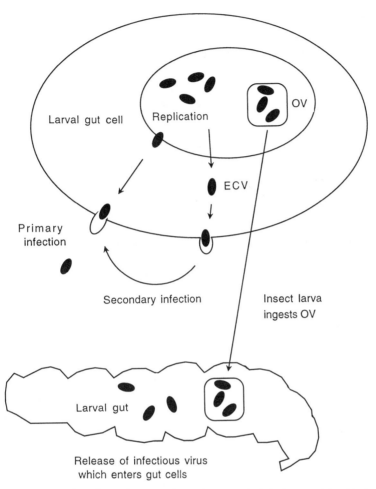

Fig. 2.1. The baculovirus life cycle. Insect larvae ingest occluded virus particles (OV), which break down in the gut to release infectious virus. These viruses enter the gut wall cells (primary infection), where they replicate to produce highly infectious extracellular virus particles (ECV), which are released into the hemolymph allowing secondary infection of cells of the gut and other tissues. OV deposited on plant leaves by decaying larvae are taken up by new hosts.

particles (ECV), and occluded virus particles (OV) that are embedded in proteinaceous viral occlusions or polyhedra composed mainly of the viral polyhedrin protein. These viral occlusions protect the virus particles from inactivation when they are exposed to the environment as dead larvae decompose. When insects feed on contaminated leaves the occlusions dissolve in the alkaline gut, releasing infectious virus. The virus lipoprotein envelope fuses with the membrane of the gut wall cells, and nucleocapsids are released into the cytoplasm. The nucleocapsids then transport the viral DNA to the cell nucleus, where replication takes place. Secondary infection of gut cells and other tissues occurs when the highly infectious ECV form of the virus buds into the hemolymph. The OV form of the virus accumulates late in infection, so that the host larva is eventually just a bag of virus, which breaks down releasing polyhedra to continue the cycle.

2. BACULOVIRUS PROPAGATION IN CULTURED CELLS

The ECV form can enter cells in culture by endocytosis, and viral occlusions are not required; nevertheless infected Sf cells produce large amounts of polyhedrin. The two major proteins involved in OV formation, polyhedrin and p10 [Vlak et al., 1988], can account for up to 50% of total cell protein; therefore, replacing the polyhedrin or p10 coding regions with a foreign gene allows the expression of very high levels of recombinant protein. Since baculovirus infection leads rapidly to the death of cells in culture, it is necessary to re-infect fresh cultures continuously. This is not ideal for commercial scale production; however, the system has many advantages for laboratory researchers. Baculoviruses only replicate in invertebrate cells, a feature that makes them one of the safest expression vectors. Infection of cultured cells can be scaled up reproducibly, the cells are easy to lyse, and expression systems are also available for secretion of protein into the medium. Our experience is that many proteins (e.g., p53) that are expressed mainly as insoluble inclusion bodies in *E. coli*, are soluble in Sf cells. Also, baculoviruses can package large amounts of DNA (probably more than 100 kbp), so the size of the inserted gene is not a restriction. Moreover, since expression from the late polyhedrin (polh) promoter occurs after maturation of infectious viral particles, expression of cytotoxic foreign proteins does not compromise virus production. Perhaps most important, however, is that proteins expressed in insect cells are subject to eukaryotic mechanisms of post-translational processing, including phosphorylation, glycosylation, fatty acid acylation, proteolytic processing, and signal peptide membrane targeting or secretion. There are, however, known to be some differences in glycosylation in comparison with vertebrate cells. Insect cells may lack some of the necessary fucose, galactose, and sialic acid transferases, and sugar side chains are usually simple and unbranched with a high mannose content. *N*-glycosylation of some proteins may also decline during late infection, so proteins expressed from the very late polh promoter may not be efficiently glycosylated, and in these cases it may be preferable to use a viral promoter that is active while cellular modification pathways are still functional (e.g., the p10 promoter, which can be active 12–32 h earlier than polh [Roelvink et al., 1992]) thus allowing a longer period for posttranslational processing.

3. INTRODUCING FOREIGN GENES INTO THE BACULOVIRUS GENOME

Since the AcMNPV genome is inconveniently large, manipulations are carried out using transplacement or transfer vectors that include the viral sequences flanking the polyhedrin (polh) gene (Fig. 2.2). The foreign gene is inserted between these sequences so that it comes under the control of the polh promoter. This whole "cassette" is then transferred to the viral genome by cotransfection of Sf cells with infectious viral DNA and the transfer vector. During the resulting infection the virus sequences flanking the foreign DNA in the transfer vector recombine with homologous sequences in the viral DNA producing recombinant virus at a low frequency. Selection of recombinant virus was originally achieved by visual detection of polyhedrin-negative viral plaques, a skilled and difficult procedure. There are now, however, several transfer vector systems that provide β-galactosidase blue/white color detection (see below). In addition, cotransfection of the transfer vector with linearized infectious viral DNA promotes recombination of homologous sequences, resulting in recovery of 30–40% recombinant virus in the first round of plaque isolation. In this case only a few viral plaques need be screened by protein or DNA analysis.

A number of commercial and noncommercial vectors are available, mostly as variations on the same basic vector. These are more extensively discussed in O'Reilly et al. [1992] and King and Possee [1992]. Our laboratories use the pVL1392 and pVL1393 vectors, which include a polylinker with several cloning sites (Fig. 2.2) and express nonfused protein. We have used these vectors in conjunction with AcRP23lacZ viral DNA [Kitts et al., 1990], which includes a unique restriction site suitable for linearizing the viral DNA, and where recombination into the viral DNA inactivates β-galactosidase so that we can detect white recombinant plaques. From different suppliers (vendors) such as Invitrogen, transfer vectors are available which express β-galactosidase simultaneously from the early to late (ETL) pro-

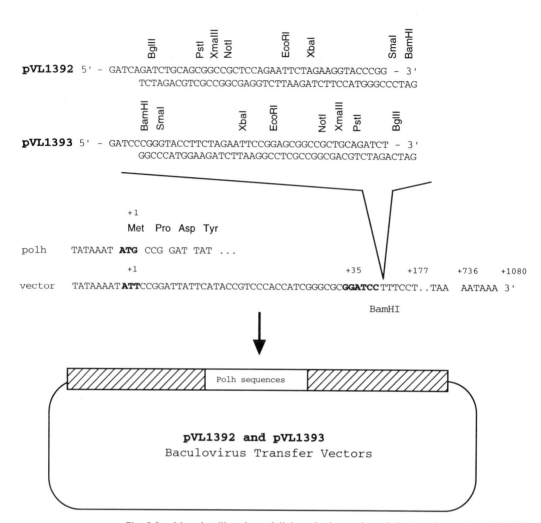

Fig. 2.2. Map detailing the polylinker cloning region of the transfer vectors pVL1392 and pVL1393 showing restriction sites available for insertion of the gene of interest. A mutation has removed the original ATG (converting it to ATT), so that translation is initiated from the ATG of the gene cloned into the polylinker 35-bp downstream.

moter. Using these allows easy selection of blue recombinant plaques. Other innovations include the expression of protein with a "tag" of histidine residues to facilitate purification by affinity chromatography, and vectors that allow secretion of protein into the medium. It is also possible to express two foreign proteins simultaneously from the same plasmid by taking advantage of the virus AcMNPV polyhedrin and p10 gene promoters. This can be more reproducible than co-infections, for example in the expression of multi-subunit proteins [Hasemann and Capra, 1990; Wang et al., 1991; Weyer and Possee, 1991].

3.1. Sf9 Cell Culture

The most efficient way to keep a stock of Sf9 cells growing continuously is to maintain them in a magnetic spinner culture flask or a flat-bottomed glass flask with a magnetic stirrer bar mixing at about 80 rpm. (Ensure that any stirrer you use is not a source of heat.) Ideally, the cells should be maintained at 27°C, but it is possible to grow them without an incubator in a room with constant temperature between 20°C and 28°C. For these media CO_2 is not required. Cells can be maintained in standard plastic tissue culture flasks without stirring, but since they will attach to the surface you will have to detach them at regular intervals for dilution. To do this, scrape the cells or knock them off with a jet of medium from a pipette or by tapping the flask. This latter method will however result in a lot of cell death since the cells attach quite tightly to plastic when grown in the presence of serum. The cells should have a population doubling time of <24 h and must be >90% viable for use in infections.

Protocol 3.1

Reagents and Materials

Aseptically prepared

- Cells: The cells we have used most extensively are Sf9 [Smith et al., 1983], (available from the American Tissue Type Culture Collection, accession number CRL-1711), but lines derived from the cabbage looper *Trichoplusia ni* (Tn368, or BTI-TN-5B1-4 also known as "High Five," available from Invitrogen), are becoming more popular because of their high-level production of recombinant proteins.
- Medium: Although a number of growth media are now available, we have had most success with EX-CELL 400 supplemented with (per 500 ml): 25 ml fetal calf serum (FCS), 5 ml of L-glutamine (200 mM, 30g/l), and 5 ml of penicillin/streptomycin solution (penicillin, 5000 U/ml, and streptomycin, 5 mg/ml). Store the medium at 4°C in the dark, and always warm to room temperature before use. Sf9 cells are very sensitive to changes in growth medium, so if you need to change to a different medium or want to use serum-free EX-CELL 400, acclimatize the cells by gradually adding in the new medium over a number of days. Bear in mind that changing the medium or serum conditions may affect protein processing (e.g., phosphorylation).
- Dimethylsulfoxide (DMSO), 10% in FCS
- Culture flasks
- Spinner flask and magnetic stirrer

Equipment
- Incubator at 27°C (CO_2 not required)

Protocol
Routine maintenance
(a) Detach cells from flask culture by scraping or dispersing with a jet of medium, or use cells grown in suspension in a spinner flask.
(b) Count by hemocytometer and determine viability by dye exclusion with trypan blue or naphthalene black.
(c) Seed spinner flask at $0.5-1 \times 10^6$ viable cells/ml (20–100 ml in a 500-ml spinner flask).
(d) Incubate at 20°C–28°C (27°C optimal).
(e) Dilute the cells to $0.5-1 \times 10^6$ cells/ml every 48–72 h, or when they have reached about $4-5 \times 10^6$ cells/ml.
(f) Transfer the cells to a clean flask every 3–4 weeks.
(g) If the cells clump try stirring them slightly faster and add the surfactant Pluronic F-68 (0.5–1.0% v/v) to reduce shearing.

Freezing cells for storage
(h) Count the cells and pellet by centrifugation at 1000 rpm (~200 g) for 5 min.
(i) Resuspend at 1×10^7 cells/ml in 10% DMSO in FCS.
(j) Aliquot into freezing tubes and chill on ice for an hour.
(k) Pack the tubes into a Styrofoam container.
(l) Freeze slowly (~1°C/min) overnight at −70°C.
(m) Transfer to liquid nitrogen freezer.
(n) To recover the frozen cells, thaw rapidly at 37°C. (If stored in the liquid phase, take care to thaw in a covered vessel to avoid risk of injury from explosion of the ampoule.)
(o) Transfer into a 25-cm^2 flask containing 5 ml of medium. Tip the flask to spread evenly.
(p) Remove the medium after 2–3 h, when most of the cells should have attached, and add 5 ml fresh medium.
(q) Leave the cells 2–3 days to recover before detaching and transferring to a stirrer flask or a larger plastic flask as described above.

3.2. Preparation of Viral DNA

We use the AcRP23LacZ virus, which carries a unique Bsu36I restriction enzyme site for linearization (see above) [Kitts et al., 1990]. More elaborate preparations for viral DNA have been described [King and Possee, 1992], but we have found the following quick procedure to be adequate.

Protocol 3.2

Reagents and Materials

Aseptically prepared
- Virus: AcRP23LacZ virus which carries a unique Bsu36I restriction enzyme site for linearization
- Medium (see Protocol 3.1)

Nonsterile
- TE: Tris HCl, 10 mM, pH 8.0, containing 1 mM EDTA
- RNAse A, DNAse free: 10 mg/ml in 10 mM Tris HCl, pH 7.5; 15 mM NaCl. Heat to 100°C for 15 min and allow to cool slowly to room temperature.
- Proteinase K dilution buffer: 0.5% SDS, 5 mM EDTA and 100 mM Tris HCl, pH 7.8
- Proteinase K: Stock, 20 mg/ml in UPW. Dilute in buffer to 50 (μg/ml for use.
- Bsu36I restriction endonuclease, for AcRP23LacZ DNA linearization
- Phenol: Saturated solution of phenol, Analytical Grade or equivalent, neutralized to pH 7.5 with 0.1 M Tris HCl, pH 7.5
- Phenol : chloroform (50 : 50)
- Chloroform, Analar or equivalent
- Sodium acetate, 3 M, pH 5.2
- Ethanol at $-20°C$

Sterile equipment
- Spinner flask

Nonsterile equipment
- Magnetic stirrer
- Centrifuge tubes for phenol extraction: 15-ml polypropylene screw cap, disposable (e.g., Falcon)
- Centrifuge tubes for pelleting virus and spinning DNA precipitate: Beckman polycarbonate tubes with caps (e.g., #355622, 10 ml, high g)
- High speed centrifuge and rotor (e.g., Beckman 45 Ti)
- Eppendorf microcentrifuge and tubes for pelleting DNA precipitate

Protocol

(a) Infect 50–250 ml of cells at 1.5–2.0 cells/ml in a stirrer (or the equivalent attached in plastic flasks) at a multiplicity of infection (moi) of 1 virus per cell (see Table 2.1 and Protocol 3.5 for virus titration).

(b) Allow the infection to proceed for 3–5 days until cell viabilty is <50% by trypan blue exclusion.

(c) Pellet the cells at 1000 rpm (~200 g) for 10 min to remove cell debris.

Table 2.1. Suggested Plating Densities for Sf9 Cell Infections

Type of Vessel	Cell Density for High-Titer Virus Stock (number of cells)	Cell Density for Protein Expression (number of cells)	Volume of Medium
24-well plate	6.0×10^5/well	1.2×10^6/well	0.5 ml
35-mm dish	1.5×10^6	3.0×10^6	1.5 ml
60-mm dish	2.5×10^6	5.0×10^6	3.0 ml
25-mm^2 flask	3.0×10^6	6.0×10^6	5.0 ml
75-mm^2 flask	9.0×10^6	1.8×10^7	10.0 ml
150-mm^2 flask	1.8×10^7	3.6×10^7	20.0 ml
Stirrer flask	$1.5–2.0 \times 10^6$/ml	$3–4 \times 10^6$/ml	As appropriate

(d) Pellet the virus particles by centrifugation of the supernatant at 35,000 rpm ($\sim$100,000 g) in a Beckman Type 45 Ti rotor or equivalent at 4°C for 30 min.

(e) Resuspend the yellowish pellet in 2 ml TE.

(f) Add RNAse A to give a concentration of 200 μg/ml and incubate at room temperature for 1 h.

(g) Digest the viral coat proteins with proteinase K, 50 μg/ml, for 2 h at 50°C.

(h) Deproteinize the viral DNA by shaking for 2 min with an equal volume of phenol in a 15 ml polypropylene tube.

(i) Collect the aqueous (upper) phase and repeat the extraction with 50:50 phenol:chloroform.

(j) Collect the aqueous (upper) layer and extract with chloroform alone.

(k) Collect the aqueous (upper) layer again and precipitate the DNA by adding 0.1 volume of 3 M sodium acetate pH 5.2, and 2 volumes of ethanol; chill at -20°C.

(l) You should see a stringy white precipitate of DNA, which you can either transfer directly to another tube, or pellet by centrifugation at 14,000 rpm ($\sim$14,000 g) in an Eppendorf microcentrifuge or equivalent.

(m) Wash once with 70% ethanol, air dry briefly and resuspend in 0.1–0.5 ml sterile TE.

(n) Store at 4°C.

Note: From 100 ml ($1.5 - 2.0 \times 10^8$) cells the yield of DNA is about 250 μg (estimated by absorbance at 260 nm).

In the case of AcRP23LacZ, linearize the viral DNA (prepared by the above procedure) prior to transfection,

(o) Digest 1 μg DNA with 5 units of Bsu36I restriction endonuclease for 2 h at 37°C in 50 μl of the manufacturers (e.g., New England Biolabs) recommended buffer.

(p) Inactivate the enzyme at 70°C for 15 min and store the digest (20 ng/μl) at 4°C.

Midgley et al.

3.3. Preparation of Transfer Vector DNA

We have used the pVL1392 and pVL1393 transfer vectors extensively (Fig. 2.2) since the multiple restriction sites usually allow single-step subcloning of foreign cDNAs. In this vector, expression is driven from the polh promoter, which reliably produces high levels of recombinant protein. Transcription is initiated from a TAAG motif located 50 nucleotides from the position of the original start codon of polyhedrin. However, there is no longer an ATG codon present and this must therefore be provided on the cDNA. Keep noncoding regions flanking your coding sequence to a minimum, as changing the spacing between the transcription start and the ATG may affect yield, although we have successfully expressed p53 from cDNAs with 135 nucleotides of intervening 5′ untranslated region. The polyhedrin leader sequence is AT-rich, so GC-rich or very long 5′ leaders should be avoided. The pVL1392, pVL1393 vector retains the polh polyadenylation region.

Protocol 3.3

Reagents and Materials
- Luria broth [Sambrook et al., 1989], 50 ml, containing 100 μg/ml ampicillin
- Qiagen tip-100 columns
- Bacterial culture (e.g., DH5, ATCC, transformed with the recombinant pVL1392, pVL1393 transfer vector or other appropriate vector)

Protocol
(a) Inoculate 50 ml Luria broth with a colony of bacteria transformed with the recombinant pVL1392, pVL1393 transfer vector or other appropriate vector, and incubate overnight shaking at 37°C.
(b) Prepare plasmid DNA using Qiagen tip-100 columns following manufacturers instructions. Any standard procedure for preparing plasmid DNA by cesium chloride density gradient centrifugation [Ausubel et al., 1996] can also be used. The circular plasmid is used directly for transfection.

3.4. Lipofectin-Mediated Transfection of Sf9 Cells

Protocol 3.4

Reagents and Materials
Aseptically prepared
- Tissue culture dishes, 35 mm
- Lipofectin

- Linearized AcRP23LacZ DNA (Bsu36I digest described in Protocol 3.2)
- Recombinant transfer plasmid (circular)
- Serum-free medium: EX-CELL 400
- Supplemented EX-CELL 400: Ex-Cell 400 supplemented with 5% FCS

Protocol

(a) Seed 1×10^6 to 1.5×10^6 cells in 35-mm tissue culture dishes and incubate for 2 h to allow them to attach.

(b) For each dish to be transfected, mix 20 μl of Lipofectin with an equal volume of a solution containing 100 ng of linearized AcRP23LacZ DNA (5 μl of the Bsu36I digest described in Protocol 3.2) and 500 ng of the recombinant transfer plasmid diluted in sterile water. Also set up an AcRP23LacZ DNA only control. Mix gently and incubate at room temperature for 15 min.

(c) Meanwhile, remove the medium from the dishes and wash twice with 2 ml of serum-free medium. Add 1.5 ml serum-free medium and leave at room temperature until the DNA incubation is ready.

(d) Add the DNA directly to the serum-free medium covering the cells. Incubate at 27°C overnight. The next day add 1.5 ml of supplemented EX-CELL 400 medium and continue incubation. At 48 h after transfection, take off and store the medium, which now contains virus particles that have budded from the infected cells.

(e) Perform a plating assay (see Protocol 3.5, below) with dilutions (10^{-1} to 10^{-3}) of the transfection supernatant medium to obtain individual viral plaques for screening.

3.5. Plating Assay to Obtain Individual Viral Plaques for Screening or Titration

Protocol 3.5

Reagents and Materials
Aseptically prepared
- Tissue culture dishes, 60 mm
- Medium: Supplemented EX-CELL 400 (as in Protocol 3.1)
- Virus supernatant from a standard infection (Protocol 3.7) or transfection (Protocol 3.4)
- Cells: Sf9
- Baculovirus agarose 2.5%: (AgarPlaque Plus Agarose) in distilled water; autoclave to sterilize. This can be remelted in a microwave oven as required.
- Neutral red: 10 mg/ml in distilled water (store at 4°C). Sterilize by filtering through a 0.2-μm filter.

- X-gal: 25 mg/ml in DMSO (store at $-20°C$)
- Constituents of first agarose overlay: 15 ml medium, 5 ml 2.5% baculovirus agarose. Melt agarose by placing in boiling water bath or by microwaving. Do not mix at this stage. Each plate will require 4 ml of the mixture, so this mix will be sufficient for 5 plates.
- Constituents of second agarose overlay: 8 ml medium, 2 ml 2.5% baculovirus agarose stock (melt agarose by placing in boiling water bath or by microwaving, do not mix at this stage), 0.1 ml neutral red (10 mg/ml), and 0.1 ml X-gal (25 mg/ml). Each plate will require 2 ml of overlay, so this will be sufficient for 5 plates.

Nonsterile equipment
- Water bath at 46°C
- Incubator at 27°C

Protocol

(a) Plate 2×10^6 cells in 60-mm tissue culture dishes, distribute evenly and leave to attach for 2 h. Try to get the cells about 50% confluent, adding more if necessary.

(b) Make the appropriate dilutions of virus in medium. Dilute transfection supernatant (Protocol 3.4) in the 10^{-1} to 10^{-3} range to obtain well-spaced plaques. To titrate virus supernatant from a standard infection (Protocol 3.7), use dilutions between 10^{-4} and 10^{-8} (virus titers should be 10^7 to 10^9 pfu/ml).

(c) Remove the medium from the attached cells and add 1 ml of diluted virus. Incubate for 1–2 h, tipping the plates occasionally to prevent them drying out.

(d) Meanwhile, prepare the first agarose overlay: Cool the agarose to 46°C for a few minutes and, at the same time, warm up the medium to 46°C. Mix in the proportions 15 ml medium to 5 ml agarose.

(e) Remove the virus from the infected cells and immediately add 4 ml of the first agarose overlay mix, making sure there are no bubbles.

(f) Allow the plates to set completely at room temperature before transferring to the 27°C incubator.

(g) Incubate for 5 days at 27°C, then overlay with 2 ml of the second agarose overlay containing neutral red dye. This accentuates the plaques which will appear as white holes in the layer of cells (Fig. 2.3D). If you are using AcRP23LacZ DNA or a transfer vector that expresses β-galactosidase, also add X-gal (5-Bromo-4-chloro-3-indolyl-β-D-galactopyranoside) to the overlay.

(h) Allow to set at room temperature, then incubate for several hours at 27°C, by which time the dyes will have diffused into the cells and you should see the plaques clearly (see Fig. 2.3D).

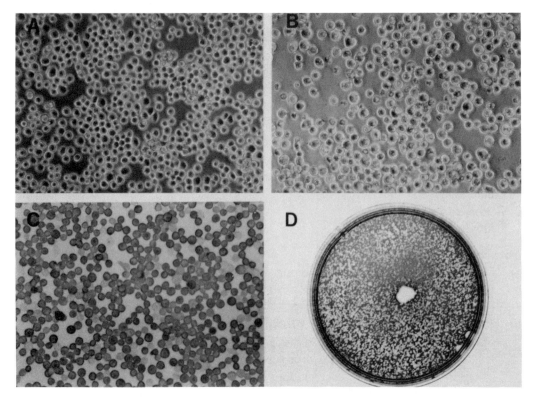

Fig. 2.3. Light microscopy of Sf9 cells. **A.** Uninfected Sf9. **B.** Sf9 5 days after infection with recombinant baculovirus expressing p53 protein. The cells become slightly swollen as their nuclei enlarge, and the debris resulting from cell lysis is clearly visible. **C.** Sf9 at 48 h postinfection with the p53 baculovirus, after fixation and staining with an antibody that specifically recognises p53 protein. **D.** A plating assay (60-mm dish) to obtain viral plaques (see Protocol 3.5), which are clearly visible after overlaying with neutral red agarose. (*A*), (*B*), and (*C*) obtained using a Zeiss Axiovert 25 microscope with 40× objective.

(i) Mark the position of plaques with a ring on the outside of the plate and continue incubating overnight.

(j) Count the viral plaques to determine titer, or go on to amplify individual viruses (Protocol 3.6).

3.6. Amplifying Recombinant Virus from Plaques for Screening

Protocol 3.6

Reagents and Materials
Aseptically prepared
- Sf9 cells
- Tips for P200

- Tissue culture plates, 24-well
- Medium: Supplemented EX-CELL 400 (as in Protocol 3.1)

Nonsterile equipment
- Micropipette Gilson P200

Nonsterile, for immunoperoxidase staining
- PBSA
- Methanol:acetone (50:50) ice cold
- Primary antibody: Antibody specific to your expressed protein at the appropriate dilution in PBSA with 10% FCS
- Secondary antibody: Appropriate peroxidase-conjugated antibody diluted 1:100 in PBSA with 10% FCS
- 3',3'-diaminobenzidine (DAB): To prepare DAB, dissolve 6 mg DAB hydrochloride in 9 ml of 50 mM Tris pH 7.6, add 1 ml of 0.3% w/v nickel chloride, then 0.1 ml 3% hydrogen peroxide. If a precipitate forms filter.

Nonsterile, for electrophoresis and western blotting
- SDS PAGE loading buffer: 50 mM Tris HCl, pH 6.8, 100 mM dithiothreitol (DTT) (add fresh just before use), 2% SDS, 0.1% bromophenol blue, 10% glycerol
- Coomassie blue (if no antibody available): 0.24 g Coomassie Brilliant Blue R250 in 90 ml methanol:H_2O (1:1 v/v) and 10 ml glacial acetic acid. Filter through Whatman No. 1 paper.
- NP40 lysis buffer: 150 mM NaCl, 1% NP40, 50 mM Tris pH 8.0, 5 mM EDTA

Protocol
(a) In 24-well tissue culture plates, Plate 10^5 Sf9 cells per well in 1 ml medium.
(b) Use a P200 Gilson micropipette and sterile tips to pick out a plug of agar over each plaque and transfer to the 24-well plate.
(c) Incubate at 27°C for 5 days.
(d) You should now see clear signs of infection: swollen cells, lysis, and debris (Fig. 2.3B).
(e) Transfer the supernatant medium, now containing virus, to sterile tubes for storage at 4°C.
    Note: This is a good opportunity to look for expression. You can use the cells remaining in the plate for immunochemical detection (Fig. 2.3C).
(f) Wash the cells once with phosphate buffered saline (PBSA) and fix for 5 min in ice cold methanol:acetone (50:50).
(g) Pour off the solution and allow the plate to dry for a few minutes.
(h) Incubate for 1 h at room temperature with primary antibody.
(i) Wash 3 times for 5 min with PBSA then incubate for 1 h at room temperature with secondary antibody.

(j)   Wash as before then develop with 3',3'-diaminobenzidine (DAB) by incubating for 5 min.

(k)   Stop the reaction by washing in water.

(l)   Examine cells for positive black staining where antibody has bound.

Western blot analysis [Sambrook et al., 1989] of infected cell lysates separated by SDS denaturing polyacrylamide electrophoresis (SDS PAGE) should then be carried out to confirm the size of the expressed protein. If you do not have an appropriate antibody, you can look for protein expression in lysates separated by SDS PAGE after staining with Coomassie blue, or label the synthesized proteins by adding [^{35}S]methionine to infected cells growing in medium minus methionine (see Section 3.9).

(n)   To lyse cells either add SDS PAGE loading buffer (50 μl/well for 24-well plate) directly to the wells after washing the cells, or lyse the cells first in NP40 lysis buffer (50 μl/well for 24-well plate)

(o)   After lysing cells, clear them of insoluble debris by centrifugation at 14,000 rpm for 10 min at 4°C in an Eppendorf centrifuge or equivalent

(p)   Add an equal volume of SDS PAGE loading buffer and perform SDS PAGE followed by Western blotting to detect protein with specific antibodies, or stain the SDS PAGE gel directly with Coomassie Blue for 1–2 h, then destain in 40% methanol, 10% acetic acid.

Always use mock infected and nonrecombinant or wild-type virus infected controls, since additional virus proteins will also be expressed during infection. When isolating a new recombinant virus, prepare virus from several different plaques derived from the original transfection, since the plaques may vary considerably in recombinant protein expression levels. You should "plaque purify" at least twice by replating dilutions of the original supernatant and repicking plaques, as described in Protocols 3.5 and 3.6, to ensure that you do not have mixed virus stocks. For subsequent infections you can also determine the virus titration the same way by counting the number of plaque-forming units in each dilution.

3.7.   Scaling Up Infection to Make High-Titer Virus Stocks

Protocol 3.7

To get efficient protein expression you will need to infect cells at a high moi of 1–10 virus particles per cell. This means that you should first make large stocks of high-titer virus supernatant by infection (see Protocol 3.2). Infections can be carried out in the appropriately sized tissue culture plate, or in a stirrer flask.

Reagents and Materials

Aseptically prepared
- Medium: Supplemented EX-CELL 400 (as in Protocol 3.1)
- Culture flasks or stirrer flask
- Sf9 cells
- Virus supernatant from Protocol 3.6

Protocol

(a) Plate out the appropriate number of cells (see Table 2.1) and add virus to a moi of 0.01 pfu/cell. The low moi at this stage reduces the frequency of further recombination events, which may create defective virus.

(b) Allow the infection to proceed for 5 days or until the majority of cells begin to lyse, then remove the debris by centrifugation at 2000 rpm ($\sim$800 g) for 10 min.

You may store the supernatant at 4°C indefinitely, but store aliquots of the virus frozen in liquid nitrogen as an emergency stock. In case of infection, the supernatant can be filtered through a 0.2 μm filter. Titrate the stock using the plating assay described in Protocol 3.5 and repeat the titration if the stock has been stored for long periods.

3.8. Infection of Cells for Recombinant Protein Expression

Protocol 3.8

Reagents and Materials

Aseptically prepared
- Culture flasks
- Cells: Sf9
- Virus supernatant
- Medium: Supplemented EX-CELL 400 (as in Protocol 3.1)
- PBSA
- Cell scraper

Protocol

(a) Plate the appropriate number of cells in flasks (see Table 2.1).

(b) Allow the cells to attach for 2 h, then remove the medium and replace with virus supernatant diluted in medium to give an moi of 1–10. You should determine the optimum infection on a small scale first. Use the minimum volume required to keep the cells covered.

(c) Allow the infection to proceed for about 2 h, then remove the supernatant, replace with fresh medium, and incubate for 24–72 h. Again you must determine the optimum period.

(d) Scrape the cells off the flasks, collect in tubes, and centrifuge at 2000 rpm (~800 g) for 10 min.

(e) Wash twice by resuspending in PBSA and repelleting.

(f) Take small samples of the cells to compare with uninfected cells by SDS PAGE (see Protocol 3.6).

(g) If you wish to store the drained cell pellets, freeze rapidly in liquid nitrogen and store at −70°C.

3.9. Metabolic Radiolabelling of Sf9 Cells with [^{35}S]Methionine

Protocol 3.9

Reagents and Materials

Aseptically prepared
- Dishes, 35 mm
- Virus supernatant
- Medium: Supplemented EX-CELL 400 (as in Protocol 3.1)
- Methionine-free medium: EX-CELL 400 without L-methionine, with added penicillin/streptomycin and 5% FCS (see Protocol 3.1.a)
- [^{35}S]methionine 370 MBq (10 mCi)/ml (e.g., Amersham SJ1015)

Protocol
(a) Plate 1.5 × 10^6 cells in 2 ml medium in a 35-mm dish and allow to attach.

(b) Remove the medium and infect with virus at a moi of 1–10 virus per cell for 1–2 h. (Optimize the infection as described in Protocol 3.8; also set up controls with no virus and with wild-type virus only.)

(c) Remove the virus and add 2 ml of medium.

(d) After incubation for 24–72 h, wash the plate with 2 ml of methionine-free medium, discard this medium, add again 2 ml of the methionine-free medium, and incubate for 30 min.

(e) Remove the medium and add 1 ml of methionine-free medium containing 15–25 μCi [^{35}S]methionine.

(f) Incubate the cells for 1–3 h.

(g) Harvest the cells for lysis and analyze total cell lysates on SDS PAGE as described in Protocol 3.6.n). Alternatively, immunoprecipitate the recombinant protein with specific antibodies before analysis on SDS PAGE.

3.10. Metabolic Radiolabelling of Sf9 Cells with [^{32}P]Orthophosphate

This labelling procedure can determine whether the protein expressed is phosphorylated in Sf9 cells. Plate out cells and infect as described in Protocol 3.9.a above.

Protocol 3.10

Reagents and Materials

Aseptically prepared

- Phosphate-free medium: phosphate free EX-CELL 400 with added penicillin/streptomycin (see Protocol 3.1.a) and 5% FCS, which has been dialyzed extensively against 0.15 M NaCl to remove phosphate.
- [^{32}P]: phosphate-free medium containing 370 KBq (10 μCi) [^{32}P]-orthophosphate, >370 MBq (>10 mCi)/ ml (e.g., Amersham PBS13)

Protocol

(a) After incubation for 24–72 h wash the dish twice with 2 ml of phosphate-free medium and discard the wash (medium).

(b) Add 2 ml of phosphate-free medium.

(c) Incubate for a further 3 h, remove the medium, and add 1 ml of phosphate-free medium containing 10 μCi ^{32}P. Enclose the dish in a Perspex box and incubate for 1–2 h.

(d) Working behind Perspex shielding, transfer the plate to ice, remove the medium, and wash twice with phosphate-free medium.

(e) Harvest the cells for lysis and analyze total cell lysates on SDS PAGE as described in Protocol 3.6.n). Alternatively, immunoprecipitate the expressed protein with specific antibodies before analysis on SDS PAGE.

4. THE USE OF RECOMBINANT BACULOVIRUS-INFECTED INSECT CELL EXPRESSION TO STUDY POST-TRANSLATIONAL MODIFICATION AND REGULATION OF ENZYME ACTIVITY

4.1 Signal Transduction Pathways Involved in Regulating Enzyme Function

Regulation of cellular growth and response to injury are partly controlled by a highly conserved network of regulatory enzymes that covalently modify, and thereby modulate, the activity of many enzymes and proteins. Such post-translational modifications include phosphorylation, glycosylation, methylation, acetylation, and prenylation. The most widely studied of these mechanisms pertaining to signal transduction are processes controlled by reversible phosphorylation [Marshall, 1994]. The high degree of evolutionary conservation of protein kinases and phosphatases and their presence in insect cells has adventitiously provided a system useful for studying post-translational regulation of enzyme function by phosphorylation. We will discuss the insights achieved in understanding the conformation and regulation of human p53 tumor suppressor protein activity using recombinant baculovirus-infected insect cell systems.

4.2. Structure-Function Relationships of the Tumor Suppressor Protein p53

4.2.1. Regulation of p53 protein conformation

The tumor suppressor p53 is a tetrameric protein [Stenger et al., 1992; Friedman et al., 1993; Wang et al., 1995] with sequence-specific DNA binding activity [El-Deiry et al., 1992] that is required for its activity as a transcription factor [Pietenpol et al., 1994]. Characterization of p53 protein has indicated that it can exist in two distinct conformations as defined by mutually exclusive binding to the monoclonal antibodies PAb1620 or PAb240 [Milner and Medcalf, 1991]. Native forms of wild type p53 bind exclusively to the antibody PAb1620, while unfolded or denatured and inactivated forms of p53 bind only to PAb240 (Fig. 2.4A). The p53 gene is frequently mutated in human tumors resulting in an unfolded and inactivated protein which binds to PAb240 [Gannon et al., 1990]. An understanding of the chemical nature of the conversion between unfolded and native p53 protein may aid the development of therapeutic strategies designed to restore wild-type conformation to a mutant p53 protein in human tumors.

Biochemical and biophysical studies of p53 require large amounts of highly purified isoforms of the protein, but the conformational flexibility of p53 tetramers has made it difficult to acquire conformational variants of the protein in a pure form. For example, expression of wild-type human p53 using established bacterial expression systems [Midgley et al., 1992], although permitting the assembly and production of a tetrameric protein [Hupp and Lane, 1994a], yields mixed forms of p53 that are partially unfolded, based on their high level of reactivity with monoclonal antibody PAb240 (Fig. 2.5A). Bacterial expression systems that produce large amounts of recombinant protein often yield aggregated complexes or inclusion bodies, and presumably the rapid synthesis of p53 and/or absence of assembly factors in bacteria prevents the efficient assembly of native tetramers. The use of insect cell expression systems has circumvented this problem by providing an excellent source of native wild-type p53 tetramers that are predominantly PAb1620-reactive (Fig. 2.5B) [Hupp and Lane, 1995]. These results demonstrate the ability of insect cell expression systems to efficiently assemble a native tetrameric form of wild-type p53.

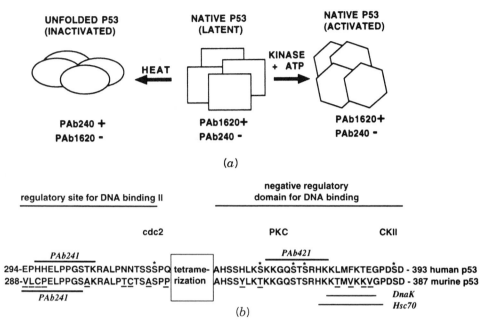

Fig. 2.4. A. Monoclonal antibody reactivity of different forms of p53 protein. p53 can be purified as a native tetramer (Hupp and Lane, 1994a) in either latent or activated forms which are both PAb1620-reactive and PAb240-nonreactive (Hupp and Lane, 1995). Activation of latent p53 can be catalyzed *in vitro* by protein kinases in an ATP-dependent reaction (Hupp and Lane, 1994b). Native p53 tetramers can be inactivated and denatured *in vitro* by thermal stress, which results in expression of the normally cryptic PAb240 epitope and reduced expression of the PAb1620 epitope (Hansen et al., 1996a). **B.** Binding site of factors that target the C-terminal regulatory domain of p53. Modification of the extreme C-terminal negative regulatory domain of p53 by casein kinase II (*), protein kinase C (*), DnaK(−), and antibody PAb421 (■) can activate p53. A second regulatory domain, containing the cdc2 phosphorylation site (*) and antibody PAb241 binding site (■), harbors a motif that cooperates with the negative regulatory domain in the activation reaction (Hansen et al., 1996b). Depicted are both human and mouse amino acid sequences in the C-terminal domain of p53, divergent sequences are underlined (_).

Further support for the use of insect cell systems to achieve the correct assembly of a conformationally sensitive protein has come from studies using mutant forms of p53. One of the most oncogenic forms of mutant p53 in human tumors is that encoded by the His175 allele [Zambetti and Levine, 1993]. The p53His175 protein derived from tumor cells is predominantly PAb240-reactive [Ory et al., 1994] and is therefore defined as being unfolded. Recent interpretations based on the crystal structure of p53 have suggested that the His175 point mutation may destabilize the packing of the core DNA binding domain and result in exposure of the normally cryptic PAb240 epitope [Vojtesek et al., 1995]. However, the p53His175 protein is predominantly PAb1620-reactive and

Fig. 2.5. Conformational integrity of purified human p53 expressed in bacterial or insect cell systems. Tetrameric forms of p53 were purified as described (Hupp and Lane, 1994a) and antibody-based two-site ELISA was used to quantitate the relative binding affinity of p53 to DO-1 (*N*-terminal monoclonal antibody), PAb1620 (monoclonal antibody specific for folded tetramers), and PAb240 (monoclonal antibody specific for unfolded or denatured p53). Relative binding affinity of the forms of p53 used in the ELISA assay: **(A)** bacterially expressed wild type human p53; **(B)** insect cell expressed wild type human p53; and **(C)** insect cell expressed mutant human p53 encoded by the His175 allele. Monoclonal antibody reactivity to p53 was determined in ELISA by quantitating the amount of peroxidase activity (O.D. 450 nm) present using an anti-rabbit IgG coupled to horseradish peroxidase.

A. Bacterially expressed human wild type p53

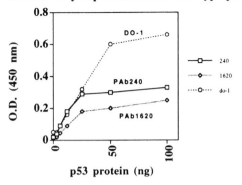

B. Insect cell expressed human wild type p53

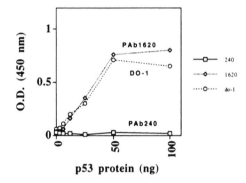

C. Insect cell expressed p53HIS175 mutant protein

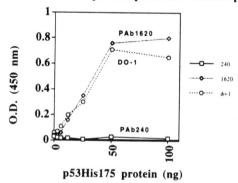

PAb240 non-reactive when expressed in insect cells (Fig. 2.5C) [Daniels and Lane, 1994]. These results indicate that the His175 point mutation does not inherently preclude the assembly of the p53His175 protein into a PAb1620-reactive tetramer and supports the use of insect cell expression systems to provide a source of native protein.

4.2.2 Regulation of p53 activity by phosphorylation

p53 protein is not only conformationally-sensitive, but it is an allosterically regulated protein, whose sequence-specific DNA binding activity can be controlled by phosphorylation or modification of a C-terminal negative regulatory domain [Hupp and Lane, 1994b; Hansen et al., 1996a,b]. One form of p53 protein is unphosphorylated in its C-terminus and is in a latent state for DNA binding, while a second form is modified in its C-terminus and is activated for DNA binding (Fig. 2.4 A and B). Both latent and activated forms of p53 are folded into the wild-type conformation, based on their exclusive reactivity to the antibody PAb1620 [Hupp and Lane, 1995]. Which enzymes catalyze p53 protein activation in cells is at present unknown, but, *in vitro*, either protein kinase C, casein kinase II, or cdc2/cyclin-dependent protein kinase family members are able to activate p53 for DNA binding. The high degree of conservation of these kinases in eucaryotes lends support to the idea of using insect cells as a model for identifying p53-activating enzymes. p53 protein is also multiply phosphorylated in insect cell expression systems, in a manner similar to that observed in human cells [Fuchs et al., 1995], and up to eleven isoforms of human p53 have been reported in insect cells [Patterson et al., 1996]. These features make this cell type an attractive system to identify kinase and phosphatase regulators of p53 function.

We have used a combined affinity/ion-exchange resin in an attempt to separate and characterize different forms of native human p53 produced in recombinant baculovirus-infected insect cells. Heparin sepharose column fractions were assayed *in vitro* either for sequence-specific DNA binding activity (Fig. 2.6) or immunochemically by Western blotting (data not shown). Using both assays, it is clear that this chromatographic method can separate at least three forms of p53; two are in an activated state and one form is latent for sequence-specific DNA binding (Fig. 2.6). Thus, insect cells can produce multiple isoforms of p53 with distinct activities, a finding that has led to identification of serum- or UV-responsive signalling pathways that can activate p53 [Hupp and Lane, 1995]. The enzymes responsible for these modifications are not known and are a current subject of investigation.

4.3. Summary

Baculovirus expression is now an accessible and user-friendly technique for the expression of large amounts of biologically active proteins. We have summarized detailed studies on the use of

Fig. 2.6. Chromatographic separation of different forms of p53 expressed in insect cells. Human p53 was expressed in Sf9 cells as described (Hupp and Lane, 1995). Lysates from infected cells were applied to a heparin-sepharose column and the bound protein was eluted with a linear salt gradient. p53 was assayed in each fraction for sequence-specific DNA binding activity using a "band-shift" assay: without antibody (lanes 1, 4, and 7); with the *N*-terminal antibody DO-1 (lanes 3, 6, and 9); or with the activating antibody PAb421 (lanes 2, 5, and 8). p53 elutes in three forms from the column: (1) activated form I, non-reactive with antibody PAb421; (2) activated form II, reactive with antibody PAb421; and (3) latent form III, inactive unless activated by PAb421. The line pointed with a closed circle marks the migration of p53-DNA complexes and the arrows mark the migration of antibody-activated or supershifted p53-DNA complexes. Copyright 1995. Used by permission of the Journal of Biological Chemistry.

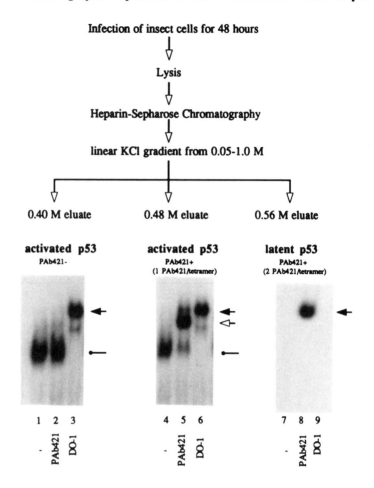

Chromatographic Separation of Three Biochemical Forms of p53

Infection of insect cells for 48 hours

Lysis

Heparin-Sepharose Chromatography

linear KCl gradient from 0.05-1.0 M

0.40 M eluate   0.48 M eluate   0.56 M eluate

activated p53   **activated p53**   **latent p53**
PAb421-   PAb421+   PAb421+
(1 PAb421/tetramer)   (2 PAb421/tetramer)

1 2 3   4 5 6   7 8 9
- PAb421 DO-1   - PAb421 DO-1   - PAb421 DO-1

baculovirus-infected recombinant insect cell expression systems to study the conformation and regulation of p53 protein. This approach is possible due to the high degree of evolutionary conservation of key enzymes that effect p53 protein function. In addition to phosphorylation, other types of conserved post-translational modifications of polypeptides have also been studied in insect cell expression systems. These include the effects of palmitylation on the activity of the dopamine receptor [Woodcock et al., 1995], regulation of glycosylation of keratins [Ku and Omary, 1994], methylation and polyisoprenylation of *ras* [Lowe et al., 1991], signal transduction cascades modulating endocytosis of CD4 [Grebenkamper and Nicolau, 1995] and the effects of phosphorylation on eIF-2 inhibition of general protein synthesis [Chefalo et al., 1994].

REFERENCES

Ausubel FM, Brent R, Kingston RE, Moore DD, Seidman JG, Smith JA, Struhl K (eds) (1996): "Current Protocols in Molecular Biology," Sections 1.0.1 to 1.8.10, New York: John Wiley & Sons.

Chefalo PJ, Yang JM, Ramaiah KV, Gehrke L, Chen JJ (1994): Inhibition of protein synthesis in insect cells by baculovirus-expressed heme-regulated eIF-2 alpha kinase. J Biol Chem 269: 25788–25794.

Daniels DA, Lane DP (1994): The characterisation of p53 binding phage isolated from phage peptide display libraries. J Mol Biol 243: 639–652.

El-Deiry WS, Kern SE, Pietenpol JA, Kinzler KW, Vogelstein B (1992): Definition of a consensus binding site for p53. Nat Genet 1: 45–49.

Friedman PN, Chen X, Bargonetti J, Prives C (1993): The p53 protein is an unusually shaped tetramer that binds directly to DNA. Proc Natl Acad Sci USA 90: 3319–3323.

Fuchs B, Hecker D, Scheidtmann KH (1995): Phosphorylation studies on rat p53 using the baculovirus expression system. Manipulation of the phosphorylation state with okadaic acid and influence on DNA binding. Eur J Biochem 228: 625–639.

Gannon JV, Greaves R, Iggo R, Lane DP (1990): Activating mutations in p53 produce a common conformational effect. A monoclonal antibody specific for the mutant form. EMBO J 9: 1595–1602.

Grebenkamper K, Nicolau C (1995): Signal transduction in SF9 insect cells: endocytosis of recombinant CD4 after phorbol ester treatment. Biochem Biophys Res Commun 207: 411–416.

Hansen S, Hupp TR, Lane DP (1996a): Allosteric regulation of the thermostability and DNA binding activity of human p53 by specific interacting proteins. J Biol Chem 271: 3917–3924.

Hansen S, Midgley CA, Lane DP, Freeman BC, Morimoto RI, Hupp TR (1996b): Modification of two distinct COOH-terminal domains is required for murine p53 activation by bacterial Hsp70. J Biol Chem 271(48): 30922–30928.

Hasemann CA, Capra JD (1990): High level production of a functional immunoglobulin heterodimer in a baculovirus expression system. Proc Natl Acad Sci USA 87: 3942–3946.

Hupp TR, Lane DP (1994a): Allosteric activation of latent p53 tetramers. Curr Biol 4: 865–875.

Hupp TR, Lane DP (1994b): Regulation of the cryptic sequence-specific DNA binding function of p53 by protein kinases. Cold Spring Harbor Symp Quant Biol 59: 195–206.

Hupp TR, Lane DP (1995): Two distinct signalling pathways activate the latent DNA binding function of p53 in a casein kinase II-independent manner. J Biol Chem 270: 18165–18174.

King LA, Possee RD (1992): "The Baculovirus Expression System. A Laboratory Guide." (London: Chapman and Hall).

Kitts PA, Ayers MD, Possee RD (1990): Linearisation of baculovirus DNA enhances the recovery of recombinant virus expression vectors. Nucleic Acids Res 18: 5667–5672.

Ku NO, Omary MB (1994): Expression, glycosylation, and phosphorylation of human keratins 8 and 18 in insect cells. Exp Cell Res 211: 24–35.

Lowe PN, Page MJ, Bradley S, Rhodes S, Sydenham M, Paterson H, Skinner RH (1991): Characterization of recombinant human Kirsten-ras (4B) p21produced at high levels in Escherichia coli and insect baculovirus expression systems. J Biol Chem 266: 1672–1678.

Lucknow VA, Summers MD (1988): Trends in the development of baculovirus expression vectors. Biotechnology 6: 47–55.

Marshall CJ (1994): MAP kinase kinase kinase, MAP kinase kinase and MAP kinase. Curr Opin Genet Dev 4: 82–89.

Midgley CA, Fisher CJ, Bartek J, Vojtesek B, Lane D, Barnes DM (1992): Analysis of p53 expression in human tumours: an antibody raised against human p53 expressed in Escherichia coli. J Cell Sci 101: 183–189.

Milner J, Medcalf EA (1991): Cotranslation of activated mutant p53 with wild type drives the wild-type p53 protein into the mutant conformation. Cell 65: 765–774.

O'Reilly DR, Miller LK, Lucknow VA (1992): "Baculovirus Expression Vectors, a Laboratory Manual." New York: WH Freeman.

Ory K, Legros Y, Auguin C, Soussi T (1994): Analysis of the most representative tumour-derived p53 mutants reveals that changes in protein conformation are not correlated with loss of transactivation or inhibition of cell proliferation. EMBO J 13: 3496–3504.

Patterson RM, He C, Selkirk JC, Merrick BA (1996): Human p53 expressed in baculovirus-infected Sf9 cells displays a two-dimensional isoform pattern identical to wild-type p53 from human cells. Arch Biochem Biophys 330: 71–79.

Pietenpol JA, Tokino T, Thiagalingam S, el Deiry WS, Kinzler KW, Vogelstein B (1994): Sequence-specific transcriptional activation is essential for growth suppression by p53. Proc Natl Acad Sci USA 91: 1998–2002.

Roelvink PW, van Meer MMM, de Kort CAD, Possee RD, Hammock BD, Vlak JM (1992): Dissimilar expression of Autographica californica multiple nulcleocapsid nuclear polyhedrosis virus polyhedrin and p10 genes. J Gen Virol 73: 1481–1489.

Sambrook J, Fritsch T, Maniatis T (1989): Molecular Cloning, a Laboratory Manual. Cold Spring Harbor, NY: Cold Spring Harbor Laboratory Press, NY.

Smith MD, Summers MD, Frazer MJ (1983): Production of human beta interferon in insect cells infected with a baculovirus expression vector. Mol Cell Biol 3: 2156–2165.

Stenger JE, Mayr GA, Mann, K, Tegtmeyer P (1992): Formation of stable p53 homotetramers and multiples of tetramers. Mol Carcinog 5: 102–106.

Vlak JM, Klinkenberg FA, Zaal KJM, Usmany M, Klinge-Roode EC, Geervliet JBF, Roosien J, Van Lent JWM (1988): Functional studies on the p10 gene of Auotographica californica nuclear polyhedrosis virus using a recombinant expressing p10 B galactosidase fusion gene. J Gen Virol 69: 765–776.

Vojtesek B, Dolezalova H, Lauerova L, Svitakova M, Havlis P, Kovarik J, Midgley CA, Lane DP (1995): Conformational changes in p53 analysed using new antibodies to the core DNA binding domain of the protein. Oncogene 10: 389–393.

Wang X, Ooi BG, Miller LK (1991): Baculovirus vectors for mutiple gene expression and for occluded virus production. Gene 100: 131–137.

Wang Y, Schwedes JF, Parks D, Mann K, Tegtmeyer P (1995): Interaction of p53 with its consensus DNA-binding site. Mol Cell Biol 15: 2157–2165.

Weyer U, and Possee RD (1991): A baculovirus dual expression vector derived from the Ac nuclear polyhedrosis virus polyhedrin and p10 promoters, co-expression of two influenza genes in insect cells. J Gen Virol 72: 2967–2974.

Woodcock C, Graber SG, Rooney BC, Strange PG (1995): Expression of the rat D2 and D3 dopamine receptors in insect cells using the baculovirus system. Biochem Soc Trans 23: S93.

Zambetti GP, Levine AJ (1993): A comparison of the biological activities of wild-type and mutant p53. FASEB J 7: 855–865.

APPENDIX: LIST OF SUPPLIERS

Material	Supplier
AgarPlaque Plus Agarose	Pharmingen
Ampicillin	Sigma
Antibodies	Dako
Bsu36I restriction endonuclease	New England Biolabs
Cell scraper	Nunc, Gibco BRL
DH5 cells	ATCC
Diaminobenzidine (DAB)	Sigma
DMSO	Sigma
EX-CELL 400	JRH Biosciences
FCS	Gibco BRL
Glutamine	Gibco BRL
Lipofectin	Gibco BRL
Magnetic stirrer	Techne
[^{35}S]Methionine (SJ1015)	Amersham
Methionine-free medium	JRH Biosciences
Neutral Red	Sigma
PBSA	Gibco BRL
Penicillin/streptomycin	Gibco BRL
[^{32}P]Phosphate (PBS13)	Amersham
Phosphate-free medium	JRH Biosciences
Plastic flasks and dishes	Nunc, Gibco BRL
Pluronic F-68	Gibco BRL
Polycarbonate tubes with caps (e.g., #355622)	Beckman
Proteinase K	Sigma
pVL1392 and pVL1393 vectors	Invitrogen
Qiagen tip-100 columns	Qiagen
RNAse A	Sigma
Sf9 cells, CRL 1711	ATCC
Spinner flasks	Techne
Tn368 cells	Invitrogen
Virus AcRP23LacZ	Gift from Kitts et al. [1990]
X-gal	Sigma

3

Electroporation of DNA into Cultured Cell Lines

Leah M. Cataldo, Zengyu Wang, and
Katya Ravid

*Boston University School of Medicine, Department of Biochemistry,
80 East Concord Street, K225 Boston, Massachusetts 02118*

DNA Transfer to Cultured Cells, Edited by Katya Ravid and R. Ian Freshney.
ISBN 0-471-16572-7 © 1998 Wiley-Liss, Inc.

I. INTRODUCTION

The ability to transfer DNA into mammalian cells *in vitro* provides a means of investigating how the overexpression of a specific gene affects cellular processes prior to advancing to more costly *in vivo* transgenic models. In addition, the regulatory domains of a specific gene's promoter region can be investigated by engineering a chimeric construct comprised of a specific gene's promoter region and a reporter gene, transferring it into mammalian cells, and measuring reporter gene expression. The methods of introducing DNA into mammalian cells are both chemical (calcium phosphate, DEAE-dextran, and lipofection) and physical (electroporation, microinjection, and particle bombardment). Because of its simplicity, efficiency, and versatility, electroporation has become a widely used technique, especially since most cells refractory to chemical methods of gene transfer are successfully transfected by electroporation [Andreason and Evans, 1988; Chu et al., 1987; Shigekawa and Dower, 1988].

1.1. Electroporation: Background and Theory

Electroporation involves subjecting a cell/DNA suspension to an electrical impulse. The electrical impulse is believed to induce local areas of reversible membrane breakdown [Zimmerman and Vienken, 1982], creating pores through which DNA enters the cell. The electroporation chamber is a cuvette with two electrodes on either side (Fig. 3.1). The cuvette is placed into the shocking chamber of an electroporation apparatus, which is charged to store energy in its capacitors. Once triggered to discharge, the electrical impulse is transmitted through the cell/DNA suspension over a period of milliseconds. In most types of electroporation apparatus, a pulse generated by the initially applied voltage decays exponentially over time. This is in contrast to apparatuses generating a square wave pulse, which is maintained at a constant voltage

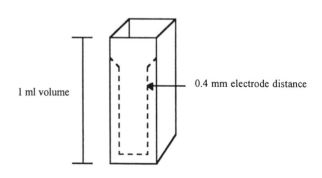

Fig. 3.1. The shocking chamber is a cuvette with noncorrosive metal electrodes along each side. The distance separating the electrodes and the total volume of the cuvette are physical parameters that should be considered when designing an electroporation protocol.

1 ml volume

0.4 mm electrode distance

Cataldo et al.

for a specific time period (Fig. 3.2). Transformation is 2- to 10-fold more efficient using an apparatus that generates an exponentially decaying pulse as opposed to a square wave pulse [Zerbib et al., 1985].

1.2. Critical Parameters for Efficient Electroporation

The parameters affecting electroporation efficiency are both physical and biological. The critical physical parameters for electroporation are field strength and pulse duration [Andreason and Evanson, 1988]. Field strength (kV/cm) depends on the voltage applied to the cells and the distance separating the electrodes of the chamber. The duration of the exponential decay depends on the amount of energy stored in the capacitors of the electroporation apparatus, measured in microfarads (μF), and the resistance of the suspension medium. A larger capacitor (150 μF) charged to the same voltage as a smaller capacitor (25 μF) will undergo a longer pulse duration because it has a greater amount of total stored energy. Field strength and pulse duration must be optimized for efficient electroporation of each individual cell type. Pilot experiments are often designed to determine optimal conditions by measuring cell viability or transient expression of a reporter gene. Electroporation is usually performed at a constant capacitance setting (and, therefore, constant pulse duration) with various field strengths (500–1500 kV/cm) for pilot investigations. For most cells, the settings at which approximately 20–50% of the cells remain viable after electroporation are sufficient for DNA transfer [Chu et al., 1987; Andreason and Evans, 1988].

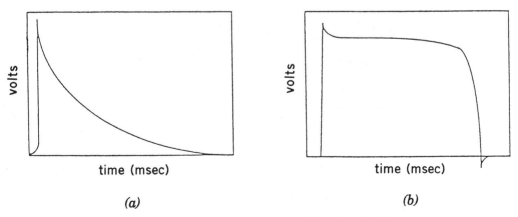

Fig. 3.2. An exponentially decaying pulse (**A**) consists of an initial voltage that decays exponentially over time. A square wave pulse (**B**) maintains the initial voltage for a specific period of time. Most commercial apparatuses generate an exponentially decaying pulse.

Temperature is also a physical parameter that may affect electroporation efficiency. For most cells, electroporation is performed at room temperature and cells are subsequently incubated on ice. The basis for placing cells on ice is that it is presumed to extend the period of time during which the membrane pores remain open [Andreason and Evans, 1989]. Performing electroporation at 4°C improves cell viability, but the effect on the efficiency of DNA transfer varies [Andreason and Evans, 1989; Potter et al., 1984].

The biological parameters that affect electroporation efficiency include the ionic strength of the cell suspension media, and the size and type of cell being transfected. Many physiological solutions, such as sucrose, impose high resistance and do not conduct electrical impulses well. The optimal solution for transmission of the electrical impulse will have low resistance and high conductivity, such as buffered saline or serum-free tissue culture medium. The presence of divalent cations in the media may affect the electroporation efficiency in some cell types [Neumann et al., 1982], and should be considered when developing an electroporation protocol. Divalent cations may stabilize the cell membrane, potentially making it more resistant to disruption by the electrical field pulse.

As mentioned previously, electroporation conditions must be optimized for each individual cell type. The nature of the cell-specific variation in buffer requirements and electroporation conditions is unknown [Andreason and Evans, 1989], but one factor may be cell diameter. In general, compared with a large-diameter cell, a small-diameter cell will survive electroporation using a larger voltage and capacitance [Chu et al., 1987; Potter, 1988], and may be used for successful electroporation protocols for various cell types.

The topology and concentration of the DNA introduced into the cell also affects the efficiency of electroporation. There is a linear relationship between DNA concentration, DNA uptake, and reporter gene expression [Boggs et al., 1986; Reiss et al., 1986; Chu et al., 1987]. It is believed that linearized DNA is more efficient for the production of stable transfectants than supercoiled DNA, presumably due to the increased efficiency with which linear DNA integrates into the genome DNA [Potter et al., 1984; Chu et al., 1987].

1.3. Advantages and Disadvantages of Electroporation over Other Methods of DNA Transfer

The advantages of electroporation over other methods of DNA transfer are that it is effective for a large number of different cell types, it is highly reproducible, and it is relatively easy to perform, since it involves fewer steps than chemical methods. Suspension cells are technically more easily transfected by electroporation than adherent cells, since adherent cells must be detached from the culture vessel. The availability of commercial electroporation apparatus facilitates determination of optimal electroporation conditions for individual cell lines, making electroporation highly reproducible. In addition, since DNA is introduced directly into the cytoplasm for translocation to the nucleus, electroporation potentially reduces the incidence of DNA mutation in some cells [Toneguzzo et al., 1988]. Chemical methods mediate DNA transfer via lysosomal compartments, potentially increasing the probability of DNA mutation or degradation by lysosomal constituents. Finally, electroporation results in the integration of DNA in low copy number [Boggs et al., 1986; Toneguzzo et al., 1988]. The copy number introduced can be adjusted by altering the concentration of DNA in the cell suspension. Chemical methods of transfection usually result in the integration of large concatamers, which may inherently interfere with cell function and obscure investigations involving specific gene overexpression [Robins et al., 1981; Kucherlapati and Skoultchi, 1984]. The drawbacks of electroporation are that it requires more cells and DNA than chemical methods of gene transfer and there is variability in optimal parameters between cell types. These disadvantages are easily surmountable, and are far outweighed by the advantages of electroporation.

2. EXPRESSION VECTORS, REAGENTS, AND SUPPLIES

2.1. Expression Vectors and Preparation of DNA

Plasmid vectors that can be engineered to contain the gene of interest are commercially available. These vectors may or may not also contain an antibiotic resistance gene to facilitate selection of positive transformants. If the vector does not contain an antibiotic resistance gene, a second vector containing an antibiotic resistance gene may be co-introduced into the cell. The molar ratio of target gene:selectable gene should be approximately 10:1 to assure sufficient copy number of target gene to achieve expression [Kucherlapati and Skoultchi, 1984]. The efficiency of electroporation may be affected by the purity of the DNA used for transfer. It is recommended that plasmid DNA be purified by cesium chloride centrifugation [Sambrook et al., 1989], phenol extraction, and ethanol precipitation. Once purified, DNA is resuspended in sterile TE (10 mM Tris HCl, 1 mM EDTA, pH 8.0). Closed circular plasmid is used for transient transfection assays; whereas for the generation of stable transfectants, DNA is linearized by restriction enzyme digestion, deproteinized by phenol extraction, ethanol precipitated, and resuspended in sterile TE.

The vectors used in the step-by-step protocols are as follows:

pCMVβ-gal [MacGregor and Caskey, 1989]
This expression vector contains the human cytomegalovirus enhancer-promoter sequences that drive the expression of the bacterial *lacz* gene encoding β-galactosidase. Expression of β-galactosidase is assayed from transfected cells by measuring enzyme activity of cell lysates in a colorimetric assay [Ravid et al., 1991a].

pSVTKGH [Selden et al., 1986]
This vector contains the SV40 enhancer and the herpes virus thymidine kinase promoter sequences driving the expression of human growth hormone (hGH). hGH is secreted directly into the tissue culture medium, and is detected by radioimmunoassay [Ravid et al., 1991b].

pcDNA3 (Invitrogen, San Diego, CA)
This expression vector contains the human cytomegalovirus enhancer-promoter sequences upstream from its multicloning site, and the polyadenylation signal and transcription-termination sequences of the bovine growth hormone gene (bGH) downstream from the multicloning site. The pcDNA3 vector also contains the gene for neomycin resistance, obviating the need to co-transfect an antibiotic resistance gene for the selection of stable transformants overexpressing a particular gene of interest. pcDNA3 may also be used as a selectable marker when co-transfected with reporter-constructs, as presented in the step-by-step protocol.

2.2. Tissue Culture Reagents, Preparation, and Supplies

The tissue culture supplies, media, and reagent preparations used in the step-by-step protocol are as follows:

Plastics
T175 flasks
6-well plates
Cell scraper

Media
Iscove's modified Dulbecco's medium (IMDM)
Phosphate buffered saline, lacking Ca^{2+} and Mg^{2+} (PBSA), $10\times$

Ham's F12 medium
Glutamine, 200 mM
Penicillin, 5000 U/ml and streptomycin, 5 mg/ml

Fetal bovine serum, heat inactivated

FBS should be stored at −20°C upon arrival from the supplier. Heat-inactivated FBS can be purchased (Gibco #16140-014), but may also be prepared by the following protocol: Completely thaw one 100-ml bottle of FBS in a 37°C water bath for 30–60 min. Stir the bottle to mix thoroughly and heat-inactivate by incubation in a 56°C water bath for 30 minutes. Aliquot 11-ml of heat-inactivated FBS into sterile 15-ml conical centrifuge tubes and store at −20°C. Aliquots should be thawed immediately before use.

Horse serum

Horse serum should be stored at −20°C upon arrival from the supplier. Completely thaw one 100-ml bottle in a 37°C water bath for 30–60 min. Mix thoroughly by stirring, aliquot 11 ml into sterile 15-ml conical centrifuge tubes, and store at −20°C. Aliquots should be thawed immediately before use.

Geneticin

Stock solutions of geneticin (G418) are prepared in PBSA, but different preparations of G418 vary in their microbiological potency. Therefore, if the potency of Lot #79P4267 is 731 μg/mg, dissolve 64.4 mg G418 per ml PBSA for a final stock solution of 50 mg/ml. Typically, 15–50 ml of stock solutions are prepared, the pH is adjusted to 7.4 and the solution is sterilized by filtration through a 0.22 μm filter. The solution is then aliquotted into sterile 1.5-ml tubes and stored at −20°C. G418 stock solutions may also be stored at 4°C.

2.3. Buffer Preparation and Apparatus for Electroporation

Electroporation buffer [Ravid et al., 1991b]

NaCl	30.8 mM
KCl	120.7 mM
Na_2HPO_4	8.1 mM
KH_2PO_4	1.46 mM
$MgCl_2$	5 mM

All reagents are ultra pure grade, purchased from American Bio-analytical (Natick, MA).

Buffered saline or serum-free media are also effective for electroporation.

Gene Pulser II apparatus
  Gene Pulser II Capacitance Extender Plus
  Gene Pulser Cuvettes

3. ELECTROPORATION OF MEGAKARYOCYTIC CELL LINES

The following protocols describe electroporation conditions for the transient transfection of the murine megakaryocytic cell line MegT [Ravid et al., 1993] and for stable transfection a clone (Y10) of the murine megakaryocytic cell line L8057 [Ishida et al., 1993]. MegT are adherent cells, which are maintained in IMDM with 4 mM glutamine and supplemented with 20% horse serum, 50 U/ml penicillin, and 50 μg/ml streptomycin. L8057-Y10 cells grow in suspension and are maintained in Ham's F12 medium supplemented with 10% heat-inactivated FBS, penicillin (50 U/ml), and streptomycin (50 μg/ml). The overall scheme presented for transient transfection of MegT cells and stable transfection of L8057-Y10 cells is similar but will be presented separately for each approach.

3.1. Transient Transfection of MegT Cells

Protocol 3.1

Reagents and Materials
Aseptically prepared
- IMDM/HS: IMDM with 4 mM glutamine and supplemented with 20% horse serum, 50 U/ml penicillin, and 50 μg/ml streptomycin
- PBSA
- Electroporation buffer
- T-75 culture flasks
- Cell scraper
- Electroporation cuvettes, prechilled
- pSVTKGH and pCMVβ-gal (circular plasmids)
Nonsterile
- Hemocytometer
- Ice bath
- Electroporation apparatus

Protocol
(a) Seed MegT cells at 1×10^5 cells/ml in a T-75 culture flask and incubate at 34°C, 5% CO_2 in IMDM/HS.
(b) During late-log phase (when cells are semi-confluent, approximately 2–3 days after seeding), remove IMDM/HS and wash adherent cells with PBSA.

(c) Add 10 ml fresh PBSA, detach the cells from the flask using a cell scraper, and collect the cells by centrifugation at 4°C, 380 g, for 5 min.

(d) Wash the cell pellet in 5 ml of electroporation buffer and count the cells using a hemocytometer. Collect the cells by centrifugation (at 4°C, 380 g, for 5 min).

(e) Resuspend the cells at 1×10^6 cells/0.8 ml in electroporation buffer. Transfer 0.8 ml of suspended cells into prechilled electroporation cuvettes.

(f) Add 10–50 μg pSVTKGH and 20 μg pCMVβ-gal (circular plasmids) to the cell suspension. When analyzing specific gene promoter-reporter gene expression, a suitable negative control is to electroporate cells in the absence of plasmid DNA.

(g) Mix the DNA/cell suspension by holding the sides of the cuvette and flicking the bottom. Incubate on ice for 10 min.

(h) Electroporate at 250 V/500 μF, record the time constant. Remove the cuvette from the shocking chamber and incubate on ice for 10 min.

(i) Transfer the electroporated cells into 10 ml IMDM/HS, rinse the cuvette with media to remove all the cells, and collect the cells by centrifugation (at 4°C, 380 g, for 5 min).

(j) Resuspend the cells in 2 ml IMDM/HS and culture in a 6-well plate at 34°C, 5% CO_2.

(k) After 3–4 days, assay for transient hGH expression in the cell supernatant and prepare a cell lysate to determine β-galactosidase activity (Section 4).

3.2. Stable Transfection of L8057-Y10 Cells

Protocol 3.2

Reagents and Materials

Aseptic
- L8057-Y10 cells
- pSVTKGH, linearized
- pcDNA3, linearized

Sterile
- F12/FBS: Ham's F12 media supplemented with 10% heat-inactivated fetal bovine serum and 50 U/ml penicillin, 50 μg/ml streptomycin
- PBSA
- Selective medium: F12/FBS containing 600 μg/ml G418
- T-75 culture flasks

Protocol
(a) Seed L8057-Y10 cells at 5×10^5 cells/ml in a T-75 culture flask and incubate at 37°C, 5% CO_2, in F12/FBS.

(b) During late-log phase, collect the cells by centrifugation at 4°C, 380 g, for 5 min.

(c) Wash the cell pellet in 10 ml PBSA and collect by centrifugation (at 4°C, 380 g, for 5 min).

(d) Wash the cell pellet in 5 ml of electroporation buffer and count the cells using a hemocytometer. Collect the cells by centrifugation as above.

(e) Resuspend the cells at 1×10^6 cells/0.8 ml in electroporation buffer. Transfer 0.8 ml of suspended cells into prechilled electroporation cuvettes.

(f) Add 50 μg of linearized pSVTKGH along with 5 μg of linearized pcDNA3 to the cell suspension. (When analyzing specific gene promoter-reporter gene expression, a suitable negative control is to electroporate cells in the absence of plasmid DNA.)

(g) Mix the DNA/cell suspension by holding the sides of the cuvette and flicking the bottom. Incubate on ice for 10 min.

(h) Electroporate at 400 V/500 μF, record the time constant. Remove the cuvette from the shocking chamber and incubate on ice for 10 min.

(i) Transfer the electroporated cells into 10 ml F12/FBS, rinse the cuvette with medium to remove all cells, and collect the cells by centrifugation.

(j) Resuspend the cells in 20 ml F12/FBS and culture in a T-75 flask at 37°C, 5% CO_2, to allow expression of the neomycin-selectable marker gene.

(k) After 24–48 h, collect the cells by centrifugation and resuspend in 20 ml of selective medium.

(l) Change selective medium every 2–4 days for at least 2 weeks to remove the debris of dead cells and to permit resistant cells to grow.

(m) Assay for hGH expression in the cell supernatant (Section 4).

4. DETECTION OF REPORTER GENE EXPRESSION

The protocols presented here employ a human growth hormone reporter gene for both transient and stable transfection. In the case of transient transfection, a plasmid containing the β-galactosidase reporter gene is co-transfected in order to normalize hGH gene expression for overall electroporation efficiency. There are commercial kits available for the detection of both hGH secretion and β-galactosidase activity (see Appendix). The hGH assay involves immunoprecipitation of hGH from cell culture medium using an ^{125}I-labeled antibody. hGH expression is quantitated using a γ-

Table 3.1. Reporter Gene Expression in MegT Cells Transfected by Electroporation

Construct Used	hGH Produced (ng/10^6 cells)	β-Galactosidase Activity (OD$_{420nm}$/10^6 cells)
pSVTKGH	7.8	—
pCMBβ-gal	—	0.51

MegT cells were transfected with 10 μg pSVTKGH and 20 μg pCMVβ-gal by electroporation, and reporter gene expression was monitored 3 days after transfection.

radiation counter. The β-galactosidase assay (see Chapters 7 and 9) involves preparing a cell lysate and measuring the enzyme activity colorimetrically with o-nitrophenyl galactopyranoside as substrate. An example of reporter gene expression in MegT cells transiently transfected by electroporation is shown in Table 3.1. There are other reporter genes that can be employed to monitor electroporation efficiency or reflect promoter activity, such as luciferase and chloramphenicol acetyltransferase (see Chapter 10). The advantages and disadvantages for each reporter gene are related to the method of detection and sensitivity of assay protocols, and the choice of reporter gene is one of personal preference and feasibility in the experimental context.

REFERENCES

Andreason GL, Evans GA (1988): Introduction and expression of DNA molecules in eukaryotic cells by electroporation. Biotechniques 6: 650–660.

Andreason GL, Evans GA (1989): Optimization of electroporation for transfection of mammalian cells. Anal Biochem 180: 269–275.

Boggs SS, Gregg RG, Borenstein N, Smithies O (1986): Efficient transformation and frequent single-site, single-copy insertion of DNA can be obtained in mouse erythroleukmia cells transformed by electroporation. Exp Hematol 14: 988–994.

Chu G, Hayakawa H, Berg P (1987): Electroporation for the efficient transfection of mammalian cells with DNA. Nucleic Acids Res 15: 1311–1326.

Ishida Y, Levin J, Baker G, Stenberg P, Yamada Y, Sasaki H, Inoue T (1993): Biological and biochemical characteristics of murine megakaryoblastic cells line L8057. Exp Hematol 21: 289–298.

Kucherlapati R, Skoultchi A (1984): Introduction of purified genes into mammalian cells. CRC Crit Rev Biochem 16: 349–379.

MacGregor GR, Caskey CT (1989): Constructs of plasmids that express E. coli β-galactosidase in mammalian cells. Nucleic Acids Res 17: 2365.

Narayanan R, Jastreboff MM, Chiu CF, Bertino JR (1986): In vivo expression of a nonselected gene transferred into murine hematopoietic stem cells by electroporation. Biochem Biophys Res Commun 141: 1018–1024.

Neumann E, Schaefer-Ridder M, Wang Y, Hofschneider PH (1982): Gene transfer into mouse myeloma cells by electroporation in high electric fields. EMBO J 1: 841–845.

Potter H (1988): Electroporation in biology: methods, applications and instrumentation. Anal Biochem 174: 361–373.

Potter H, Weir L, Leder P (1984): Enhancer-dependent expression of human k immunoglobulin genes introduced into mouse pre-B lymphocytes by electroporation. Proc Natl Acad Sci USA 81: 7161–7165.

Ravid K, Beeler DL, Rabin MS, Ruley HE, Rosenberg RD (1991a): Selective targeting of gene products with the megakaryocyte platelet factor 4 promoter (in transgenic mice). Proc Natl Acad Sci USA 88: 1521–1525.

Ravid K, Doi T, Beeler DL, Kuter DJ, Rosenberg RD (1991b): Transcriptional regulation of the rat platelet factor 4 gene: interaction between an enhancer/ silencer domain and the GATA site. Mol Cell Biol 11: 6116–6127.

Ravid K, Li YC, Rayburn HB, Rosenberg RD (1993): Targeted expression of a conditional oncogene in hematopoietic cells of transgenic mice. J Cell Biol 123: 1545–1553.

Reiss M, Jastreboff MM, Bertino JR, Narayanan R (1986): DNA-mediated gene transfer into epidermal cells using electroporation. Biochem Biophys Res Commun 137: 244–249.

Robins DM, Ripley S, Henderson AS, Axel R (1981): Transforming DNA integrates into the host chromosome. Cell 23: 29–39.

Selden RF, Howie KB, Rowe ME, Goodman HM, Moore D (1986): Human growth hormone as a reporter gene in regulation studies employing transient gene expression. Mol Cell Biol 6: 3173–3179.

Sambrook J, Fritsch EF, Maniatis T (1989): "Molecular Cloning: A Laboratory Manual," 2nd ed. Plainview, NY: Cold Spring Harbor Laboratory Press.

Tonneguzo F, Keating A, Glynn S, McDonald K (1988): Electrical field mediated gene transfer: characterization of DNA transfer and patterns of integration in lymphoid cells. Nucleic Acids Res 16: 5155–5532.

Shigekawa K, Dower WJ (1988): Electroporation of eukaryotes and prokaryotes: a general approach to the introduction of macromolecules into cells. Biotechniques 6: 742–751.

Zerbib D, Amalric F, Teissie J (1985): Electric field mediated transformation: isolation and characterization of TK⁺ subclone. Biochem Biophys Res Commun 129: 611–618.

Zimmermann U, Vienken J (1982): Electric field-induced cell-to-cell fusion. J Membr Biol 67: 165–182.

APPENDIX: LIST OF SUPPLIERS

Item	Catalogue #	Supplier
Cell scraper	08-733-2	Fisher Scientific
Fetal bovine serum, FBS	26140-012	Gibco
Fetal bovine serum, FBS, heat inactivated	16140-014	Gibco
Gene Pulser Cuvettes	165-2088	BioRad
Gene Pulser II Apparatus	165-2105	BioRad
Gene Pulser II Capacitance Extender Plus	165-2108	BioRad
Geneticin, G418-sulfate	11811	Gibco
Glutamine, 200 mM	25030-016	Gibco
Ham's F12 medium	11765-021	Gibco
hGH secretion detection kit	40-2205	Nichols Institute Diagnostics
Horse serum	16050-015	Gibco
Iscove's modified Dulbecco's medium IMDM	12440-020	Gibco
KCl	AB 1652	American Bioanalytical

APPENDIX (*Continued*)

KH_2PO_4	AB 1650	American Bioanalytical
$MgCl_2$	AB 1310	American Bioanalytical
Na_2HPO_4	AB 2060	American Bioanalytical
NaCl	AB 1915	American Bioanalytical
Penicillin, 5000 U/ml and streptomycin, 5 mg/ml ($100\times$)	15070-063	Gibco
Phosphate buffered saline, $10\times$ PBSA	70011-036	Gibco
Plasmid, pcDNA3		Invitrogen
Plates, 6-well	62406-161	VWR
β-galactosidase activity detection kit	E2000	Promega
T75 flasks	29442-034	VWR

4

DNA Transfer to Blastula-Derived Cultures from Zebrafish

Paul Collodi

Department of Animal Sciences, Purdue University, West Lafayette, Indiana 47907

DNA Transfer to Cultured Cells, Edited by Katya Ravid and R. Ian Freshney.
ISBN 0-471-16572-7   © 1998 Wiley-Liss, Inc.

I. INTRODUCTION

The zebrafish possesses many favorable characteristics that make it a popular nonmammalian model for studies of vertebrate development and toxicology [Powers, 1989; Rossant and Hopkins, 1992; Driever et al., 1994; Nusslein-Volhard, 1994]. Zebrafish reach sexual maturity in approximately 3 months and females produce 100 to 200 eggs each week throughout the year. Embryogenesis is completed outside of the mother in 3 to 4 days and the large, transparent embryos are amenable to experimental manipulations involving cell labeling or ablation techniques [Westerfield, 1993; Lin et al., 1992; Ho and Kimmel, 1993] (Fig. 4.1). Genetic approaches can be employed for the study of zebrafish development using methods that have been established for the production of haploid and homozygous diploid embryos [Streisinger et al., 1981; Westerfield, 1993]. These methods have enabled researchers to isolate several mutant lines of fish and identify genes essential for embryonic development [Mullins and Nusslein-Volhard, 1993; Gaiano et al., 1996b]. Transgenic lines of zebrafish have been produced by introduction of plasmid DNA into embryos at the one- or two-cell stage and of viral DNA into the blastula [Stuart et al., 1988, 1990; Culp et al., 1991; Bayer and Campos-Ortega, 1992; Lin et al., 1994a,b; Gaiano et al., 1996a,b].

In vitro approaches using embryo cell cultures have also been employed for the of study zebrafish development. Established cell lines and long-term primary cultures, initiated from blastula, gastrula, and late-stage embryos [Collodi et al., 1992a,b; Ghosh and Collodi, 1994; Sun et al., 1995a,b; Peppelenbosch et al., 1995; Driever and Rangini, 1993] have been used for gene transfer experiments to identify and characterize genetic regulatory sequences that are functional in the zebrafish [Sharps et al., 1992; Driever and Rangini, 1993], and for the study of extracellular factors controlling cell differentiation and gene expression [Bradford et al., 1994a; Wu et al., 1997]. Cultured embryo cells have also been utilized for the production of chimeric fish [Collodi et al., 1992b; Sun et al., 1995b].

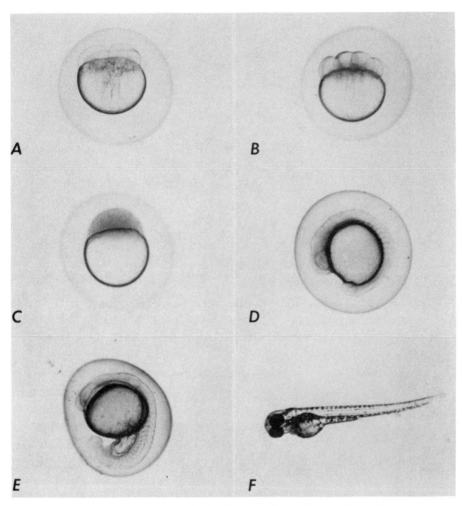

Fig. 4.1. Stages of zebrafish embryonic development. **A.** 4-cell stage. **B.** 8-cell stage. **C.** blastula. **D.** 14-somite stage. **E.** 18-somite stage. **F.** 2-day-old hatched fish.

2. ZEBRAFISH EMBRYO CELL CULTURES

2.1. Cell Lines Derived from Early-Stage Embryos

Since differentiation occurs during zebrafish gastrulation, the cells in earlier stage embryos, such as the blastula, are pluripotent [Lin et al., 1992; Kane et al., 1992; Ho and Kimmel, 1993], and methods have been developed for the culture of cells from these early-stage embryos. The ZEM-2 cell line, initiated from mid-blastula-stage embryos, has been growing in culture for more than 300 generations in medium containing low concentrations of fetal bovine and trout sera, insulin, trout embryo extract and in medium conditioned by buffalo rat liver cells (Fig. 4.2) [Ghosh and Collodi,

Fig. 4.2. A. ZEM-2 cells cultured in LDF medium supplemented with insulin (10 μg/ml), trout serum (0.5%), trout embryo extract (40 μg/ml), and FBS (1%). Cultures grown 3 days in the same medium supplemented with **B.** 0.5 μg/ml retinoic acid, or **C.** 10 ng/ml TGF-β.

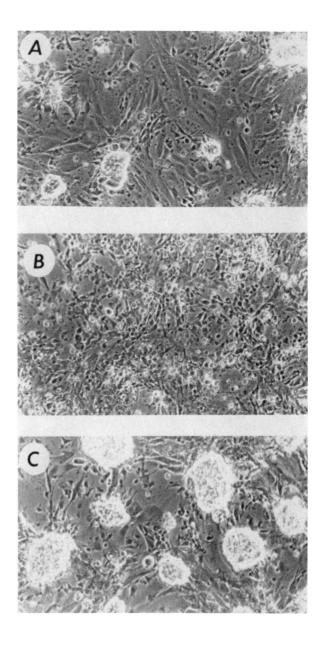

1994; Wu et al., 1997]. When introduced into a host embryo, ZEM-2 cells contribute to tissues derived from all three germ layers [Speksnijder et al., 1997] and in culture the cells exhibit a mitogenic response to several purified mammalian peptide growth factors including TGF-β, FGF, and PDGF [Wu et al., 1997]. In addition to influencing ZEM-2 cell proliferation, TGF-β and retinoic acid regulate extracellular matrix (ECM) protein synthesis in the cultures, and the presence of TGF-β and retinoic acid results in dramatic changes in cell morphology (Fig. 4.2). A second line, ZEM-2A, was derived from ZEM-2 by selecting cells able to grow in a simplified medium supplemented with FBS [Ghosh and Collodi, 1994]. The growth rates of the two cell lines are similar; however, unlike the parent line, ZEM-2A cells do not respond morphologically to TGF-β or retinoic acid.

Another blastula-derived cell culture, ZEMH, was initiated from haploid zebrafish embryos, which were developed from eggs fertilized with UV-inactivated sperm [Collodi et al., 1992a]. ZEMH was derived by the same procedures that were used for the derivation of ZEM-2 and has been propagated for more than 30 generations. Karyotype analysis has revealed that the ZEMH culture consists of a large percentage of cells containing the haploid chromosome number. The cells exhibit inducible levels of cytochrome P450 enzyme activity and spontaneously differentiate to a pigmented phenotype.

A fibroblastic cell line, ZEF, has also been derived from early-stage embryos in medium supplemented with FBS and FGF. Once established, the line has been maintained in LDF medium (see below) containing 10% FBS [Sun et al., 1995a], and ZEF cells have been used as a feeder layer for primary cultures of zebrafish embryo cells [Sun et al. 1995a; Bradford et al., 1994a].

2.2. Primary Cultures Derived from Early-Stage Embryos

In addition to the continuously growing embryo cell lines that are available, methods have been developed for the initiation of primary cultures derived from early zebrafish embryos [Sun et al., 1995b; Bradford et al., 1994a,b; Gaiano et al., 1995; Furutani-Seiki et al., 1995]. The primary cultures were initiated in conditions that are similar to those used for the derivation of mouse ES cell lines [Evans and Kaufman, 1981; Martin, 1981], and the fact that zebrafish ES-like cultures could be maintained for several weeks made them suitable for many types of gene transfer studies [Sun et al., 1995b].

Primary cultures, derived from early gastrula-stage embryos, maintained a diploid chromosome number and exhibited a morphology characteristic of pluripotent ES cells when derived on a feeder layer of ZEF fibroblasts in medium containing FBS, trout serum, fish embryo extract, insulin, and leukemia inhibitory factor [Sun et al. 1995b; Bradford et al., 1994a]. The primary cultures also exhibited other characteristics of pluripotent cells, including an elevated level of alkaline phosphatase activity and the ability to participate in normal embryonic development when introduced into host embryos. In the absence of a feeder layer, the cells underwent melanocyte differentiation, which was inhibited by the addition of bFGF to the medium [Bradford et al., 1994a]. When the embryo cells were cultured on polylysine substrate in the absence of a feeder layer, they differentiated into neuronal cell types, as evidenced by the formation of axon and dendrite-like processes (Fig. 4.3), and expression of the neural markers GFAP and neurofilament protein [Sun et al., 1995b; Ghosh et al., 1997]. Under these conditions the cultures also exhibited elevated levels of acetylcholinesterase activity that was inhibited by carbaryl and malathion [Ghosh et al., 1997].

2.3. Cell Lines Derived From Late-Stage Embryos

Late-stage zebrafish embryos (20–24 h post-fertilization) have been utilized for the derivation of three fibroblastic cell lines, ZF29, ZF13 [Peppelenbosch et al., 1995], and ZF4 [Driever and Rangini, 1993]. The lines were derived in Leibowitz's L-15 medium (ZF29 and ZF13), or a mixture of Ham's F12 and Dulbecco's modified Eagle's media (ZF4) supplemented with FBS. The ZF29

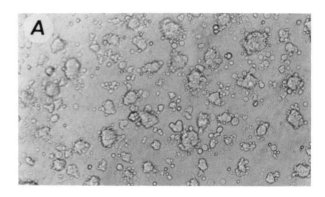

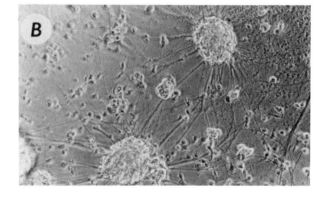

Fig. 4.3. A. Primary cultures of zebrafish blastula-derived embryo cells grown for 3 days on polylysine-coated plates in LDF medium supplemented with BRL-conditioned medium (50%), FBS (1%), insulin (10 μg/ml), trout serum (0.5%), and trout embryo extract (40 μg/ml). **B.** Ten-day-old culture showing the network of neurites extending from each cell aggregate.

and ZF13 cell lines have been utilized for the study of growth factor–induced ionic signaling. Also, microinjection experiments have demonstrated that ZF29 cells differentiate into mesodermal tissue *in vivo* and are capable of inducing ectopic notochord and neural tissue formation in host embryos [Peppelenbosch et al., 1995; Speksnijder et al., 1997].

3. DNA TRANSFER TO ZEBRAFISH EMBRYO CELLS

Three methods have been used to introduce exogenous DNA into embryo cell cultures from zebrafish and other fish species. Plasmid DNA has been transferred to the cultured cells using transfection mediated by both calcium phosphate precipitate and liposomes [Helmrich et al., 1988; Sharps et al., 1992; Fu and Aoki, 1991; Collodi et al., 1992a; Dreiver and Rangini, 1993], and viral DNA has been introduced by retrovirus infection [Burns et al., 1993; Lin et al., 1994a]. Colonies of stably transfected zebrafish embryo cells have been isolated by introduction of the aminoglycoside phosphotransferase (*neo*) selectable marker gene followed by the addition of the neomycin analogue, G418, to the medium. Resistant colonies have been expanded to produce permanent transfected lines [Collodi et al., 1992a], and integration and expression of *neo* has been used as an assay system to evaluate the gene-trapping ability of various plasmid constructs [Culp et al., 1995].

Transient expression of marker genes such as *lacZ* and *CAT* have been used to identify promoter sequences and to determine the relative strength of various promoter/enhancers in zebrafish cultures [Sharps et al., 1992; Fu and Aoki, 1991; Driever and Rangini, 1993]. Results of this work have shown that several promoters obtained from viruses, mammals and fish are all active in the zebrafish cultures but the strongest expression is obtained with the viral promoters. High levels of expression were achieved in blastula (ZEM-2) and late-stage embryo (ZF4) cell lines using the CMV immediate early promoter/enhancer [Sharps et al., 1992; Driever and Rangini, 1993]. Strong expression was also obtained with RSV- (ZF4) and SV40- (ZEM-2) derived promoters.

In addition to the plasmids that have been used for gene transfer studies with embryo cells in culture, several other DNA constructs containing a variety of promoter/enhancers and reporter genes have been introduced and expressed in fish embryos [Stuart et al., 1988, 1990; Bayer and Campos-Ortega, 1992; Amsterdam et al., 1995, 1996; Lin et al., 1994b; Gibbs et al., 1994]. In one study, incorporation of transgenic DNA into the embryo was enhanced in the presence of MoMLV integration protein [Ivics et al., 1993]. Two laboratories have used vectors containing regulatory sequences derived from fish β-actin to produce transgenic zebrafish, goldfish, and medaka [Moav et al., 1993; Caldovic and Hackett, 1995; Takagi et al., 1994]. Introduction of vectors containing carp β-actin promoter/enhancer resulted in a high level of transgene expression in goldfish and lower levels in zebrafish. Expression of these vectors in zebrafish was influenced by interactions between regulatory elements contained in the proximal promoter and the first intron [Moav et al., 1993]. Utilization of the medaka β-actin promoter resulted in high levels of transgene expression in medaka, but the construct was not tested in zebrafish [Takagi et al., 1994]. Amsterdam et al. [1995, 1996] used the Xenopus elongation factor 1α promoter/enhancer to drive transgene expression in zebrafish. Using this promoter they demonstrated that the green fluorescent protein (GFP) from the cnidarian *Aequorea victoria* can be employed as a reporter in live embryos. Fluorescence from GFP can be detected without an exogenously added substrate. Muscle-specific expression of GFP was obtained in transgenic zebrafish containing the reporter gene under the control of rat myosin light chain enhancer [Moss et al., 1996]. In addition to GFP, *lacZ* and luciferase genes have also been employed as reporters in live zebrafish embryos. Lin et al. [1994b] demonstrated that *lacZ* expression can be visualized in the embryos following a 2–3 min incubation in the substrate FDG, and expression of luciferase has been observed using a film assay for luminescence [Gibbs et al., 1994].

In addition to transfection and microinjection, viral infection has also been used to deliver foreign DNA to the genome of zebrafish embryo cells in culture and to whole embryos [Burns et al., 1993; Lin et al., 1994a, Gaiano et al., 1996a,b]. Burns et al. [1993] produced a pseudotype of MoMLV that was able to infect the ZF4 zebrafish embryo cell line as well as cells from other fish species. The genome of the pseudotype virus, derived from MoMLV, was encapsidated in an envelope containing the G glycoprotein of VSV in place of the retroviral *env* glycoprotein. Encapsidation of the MoMLV genome in the VSV envelope gave the pseudotype virus the broad host range that is characteristic of VSV. Two other important advantages of the pseudotype virus are that it can be concentrated to high titers by ultracentrifugation with a minimal loss of infectivity, and that it can integrate into the germline when used as vector to produce transgenic zebrafish [Hopkins, 1993; Lin et al., 1994a]. When the virus was injected into zebrafish blastulas, 16% of the embryos transmitted proviral insertions to their offspring. The frequency increased to 83% when two new MoMLV-based retroviral vectors were used [Gaiano et al., 1996a,b]. Also, the percentage of offspring containing proviral insertions and the number of insertions incorporated increased dramatically with the new pseudotype viruses. This advancement has enabled researchers to generate large numbers of transgenic zebrafish and pursue a strategy involving large-scale insertional mutagenesis to identify genes that are essential for embryonic development [Gaiano et al., 1996b]. Research involving insertional mutagenesis and gene tagging to study zebrafish development may be further advanced by the characterization of a family of tRNA-derived short, interspersed DNA elements that are unique to the genus *Danio* [Izsvak et al., 1995, 1996; Ivics et al., 1996].

The following sections describe materials and methods for the culture of zebrafish embryo cells, transfer of DNA to the cultures, and selection of stable transfectants. Use of the embryo cell cultures for the production of chimeric fish is also discussed.

4. PREPARATION OF MEDIA AND REAGENTS

4.1. LDF Medium

Cultures derived from early-stage zebrafish embryos are initiated and maintained in LDF medium, which is a mixture of Leibowitz's L-15 medium, Dulbecco's modified Eagle's medium, and Ham's F12 medium. One liter of each medium is prepared separately, along with stock solutions of HEPES buffer and each antibiotic. The components are then mixed in the proper ratio to prepare LDF. High-purity deionized, distilled water (UPW) is used for the preparation of the medium and other reagents.

HEPES buffer (100×)

Dissolve 178.7 g of HEPES in 500 ml of UPW, adjust the pH to 7.4, and store frozen in 20-ml aliquots (−20°C).

Penicillin G (100×)

Dissolve 6 g of penicillin G in 500 ml of UPW and store frozen in 20-ml aliquots (−20°).

Streptomycin (100×)

Dissolve 10 g of streptomycin sulfate in 500 ml of UPW and store frozen in 20-ml aliquots (−20°C).

Ampicillin (100×)

Dissolve 1.25 g of ampicillin in 500 ml of UPW and store frozen in 20-ml aliquots (−20°C).

Leibowitz's L-15 medium

Add the contents of one packet of powdered L-15 medium to 500 ml of UPW in a 1-l-Erlenmeyer flask and add 10 ml each of HEPES (100×), penicillin (100×), streptomycin (100×), and ampicillin (100×). Stir until the powder is completely dissolved and adjust the volume to 1 liter. Filter-sterilize (0.2-μm membrane) and store refrigerated (4°C) in 200-ml aliquots.

Dulbecco's modified Eagle's (DMEM) medium

Add the contents of one packet of powdered DMEM to 500 ml of UPW in a 1-l-Erlenmeyer flask and add 10 ml each of HEPES (100×), penicillin (100×), streptomycin (100×), and ampicillin (100×). Stir until the powder is completely dissolved and adjust the volume to 1 liter. Filter-sterilize (0.2-μm membrane) and store refrigerated (4°C) in 200-ml aliquots.

Ham's F-12 medium

Add the contents of one packet of powdered F-12 to 500 ml of UPW in a 1-l-Erlenmeyer flask and add 10 ml each of HEPES (100×), penicillin (100×), streptomycin (100×) and ampicillin (100×). Stir until the powder is completely dissolved and add 1.2 g of $NaHCO_3$. Adjust the volume to 1 liter, filter-sterilize (0.2-μm membrane), and store frozen (−20°C) in 200-ml aliquots. After thawing, store the medium refrigerated at 4°C.

Selenite

To prepare a concentrated stock solution, dissolve 108 mg of sodium selenite in 100 ml of UPW and store frozen (−20°C). To prepare a working stock solution, dilute the concentrated stock 1:1000 with UPW, and filter-sterilize (0.2-μm membrane). Store as 5-ml aliquots.

LDF basal nutrient medium

To make 200 ml of LDF, mix:

Leibowitz's L-15 medium ...... 100 ml
DMEM .... 70 ml
Ham's F-12 medium .... 30 ml
Stock selenite solution .... 200 μl

Store medium refrigerated at 4°C.

LDF primary medium

FBS 1%
Trout serum 0.5%

Trout embryo extract .... 40 μg protein/ml
Insulin .... 10 μg/ml
Leukemia inhibitory factor, human,
recombinant .... 10 ng/ml

   In LDF basal medium

LDF maintenance medium
FBS 1%
Trout serum 0.5%
Trout embryo extract .... 40 μg protein/ml
Insulin .... 10 μg/ml
BRL-conditioned medium 50%

   In LDF basal medium

LDF selection medium
LDF maintenance medium containing
Geneticin (G418) .... 500 μg/ml

4.2.   Other Cell Culture Reagents

Insulin

   Dissolve 250 mg of insulin in 250 ml of 20 mM HCl. Filter-sterilize (0.2-μm membrane) and store 4-ml aliquots frozen ($-20°$C) in sterile polypropylene tubes. After thawing, store refrigerated at 4°C.

Phosphate-buffered saline (PBSA)

   Dissolve 0.2 g of KCl, 0.2 g of KH_2PO_4, 8 g of NaCl and 2.16 g of $Na_2HPO_4 \cdot 7H_2O$ in 800 ml of UPW and adjust the pH to 7.0. Adjust the final volume to 1 liter, sterile-filter (0.2-μm membrane) and store refrigerated at 4°C.

Trout serum

   Trout serum can be obtained commercially from East Coast Biologicals (Sea Grow), or it may be prepared in the laboratory from whole trout blood. To prepare serum from whole trout blood, allow the blood to clot overnight (at 4°C) and remove cells and clotted protein by centrifugation. Remove the serum supernatant, filter-sterilize (0.2-μm membrane) and heat inactivate by incubating at 56°C for 20 min. Store the heat-inactivated serum frozen ($-80°$C) in 1- to 2-ml aliquots.

Trout embryo extract

   Collect embryos (Shasta Rainbow, other strains of trout 28 days after fertilization, reared at 10°C, or zebrafish 3 days after fertilization, reared at 28°C) and store frozen at $-80°$C. To prepare the extract, thaw the embryos (approximately 150 g) and homogenize in 10 ml of LDF for 2 min on ice, using a Tissuemizer homogenizer (Tekmar). Pass the homogenate through several layers of cheesecloth to remove the chorions and then centrifuge (20,000 g, 30 min, at 4°C). After centrifugation, remove the supernatant, leaving behind the bright orange lipid layer present on the surface. Transfer the supernatant to a new tube and centrifuge a second time as before. Following the second low-speed centrifugation, remove the supernatant, leaving behind any remaining lipid, and then ultracentrifuge (100,000 g, 60 min, at 4°C). After ultracentrifugation, collect the supernatant, leaving behind the lipid layer, dilute with LDF (1:10), and filter-sterilize. Due to its viscosity, the extract must be passed through a series of filters (1.2-μm, 0.45-μm, and 0.2-μm). Store the extract frozen (−80°C) in 0.5-ml aliquots. To use the extract for cell culture, measure the protein concentration [Bradford, 1976] and then dilute the extract with LDF to the desired working concentration. Store the dilute extract refrigerated (4°C) for a maximum of 2 months.

BRL cell-conditioned medium

Culture BRL cells (ATCC) at 37°C in 75-cm^2 flasks in FD medium (1:1 mixture of Ham's F-12 medium and Dulbecco's modified Eagles medium) supplemented with 10% FBS. When the cultures become confluent, replace the FD medium with LDF supplemented with 2% FBS, and incubate the cells at 37°C with the cap screwed down tight on the flask. After 5 days, remove the LDF, filter, and store frozen (−20°C). Add fresh LDF to the BRL cultures and repeat the process 5 days later. Conditioned LDF medium can be collected up to three times from the same flask before the cells must be split and allowed to grow again to confluency.

Holtfreter's buffer

NaCl .... 70 g
KCl .... 1.0 g
NaHCO$_3$ .... 4.0 g
CaCl$_2$ .... 2.0 g
UPW to give .... 1000 ml

Store at 4°C.
Prepare a working solution by diluting 1:20 with UPW.

Pronase E solution (10×)

Dissolve 250 mg of pronase E powder in 50-ml of Holtfreter's buffer and filter-sterilize (0.2-μM membrane). Prepare a working solution (0.5 mg/ml) by diluting 1:10 with sterile Holtfreter's buffer. Store 10-ml aliquots frozen at −20°C.

Trypsin

Dissolve 1 g of trypsin (cell culture grade) and 0.186 g of EDTA in 500 ml of PBSA. Filter-sterilize (0.2-μm membrane) and store frozen ($-20°C$) in 20-ml aliquots.

Feeder layers

Feeder layers of embryonic fibroblasts are prepared from gastrula-stage zebrafish [Sun et al., 1995a] (see also Protocol 5.1.1).

(1) Collect approximately 30 embryos (8 h after fertilization).

(2) Remove the chorion by pronase treatment [Sun et al., 1995a]

(3) Incubate the embryos in trypsin (1 min), while gently pipetting to dissociate the cells.

(4) Add FBS (10% final concentration) to stop the action of the trypsin.

(5) Collect the cell by centrifugation (500 g, 10 min).

(6) Resuspend the cell pellet in 5 ml of LDF primary medium containing bovine fibroblastic growth factor (10 ng/ml) and transfer the cells to a 25-cm^2 flask.

(7) Allow the cells to attach and grow to confluency at 26°C.

(8) When the cells are confluent, passage the culture in the same medium.

(9) After 2–3 passages, a mixed population of epithelial and fibroblastic cells are present in the culture, and the cells can be maintained in LDF containing only 10% FBS. Select the fibroblasts for further culture, by differential trypsinization.

> (a) Treat the culture with trypsin for 1 min to remove most of the fibroblasts, and leave the epithelial cells attached to the plastic.
>
> (b) Transfer the fibroblasts to another flask, and repeat this process when the culture becomes confluent.
>
> (c) After 2 or 3 passages, the culture will consist of a homogeneous population of fibroblasts.

(10) To prepare feeder layers of growth-arrested fibroblasts, add the cells to the appropriate culture dish or flask and allow the fibroblasts to grow into a confluent monolayer.

(11) Add mitomycin C, (10 μg/ml), to the cultures and incubate for 3 h at 26°C.

(12) After rinsing 3× with LDF, the growth-arrested fibroblasts can be used as feeder layers for zebrafish embryo cell cultures.

4.3.   Reagents for Calcium Phosphate Transfection

Calcium chloride solution (2.5 M Stock)

Dissolve 7.35 g $CaCl_2$ in 20 ml UPW and filter-sterilize (0.2-μm membrane). Store frozen at $-20°C$.

HEPES-buffered saline

Dissolve 730 mg NaCl, 595 mg HEPES, and 13 mg Na_2HPO_4 in 50 ml UPW. Adjust pH to 7.0 and filter-sterilize (0.2-μm membrane). Store frozen at $-20°C$.

4.4. Reagents for *lacZ* Staining

Fixative solution

Formaldehyde, 37.1% commercial stock solution	5.2 ml
Glutaraldehyde, 25% commercial stock solution	0.8 ml
PBSA	94 ml

Final concentration is 2% formaldehyde, 0.2% glutaraldehyde.

4-Cl-5-Br-3-indolyl-β-galactoside (X-Gal)

To prepare a 50× stock solution, dissolve 50 mg X-Gal in 1ml DMSO. Wrap the tube in foil to protect from light exposure and store at $-20°C$.

Staining Buffer

Dissolve 165 mg potassium ferricyanide $[K_3Fe(CN)_6]$, 211 mg potassium ferrocyanide $[K_4Fe(CN)_6]$, and 41 mg magnesium chloride $[MgCl_2]$ in 100 ml PBSA.

5. DNA TRANSFER TO ZEBRAFISH EMBRYO CELLS

5.1. Embryo Cell Culture

5.1.1. Primary cultures

Protocol 5.1.1

Reagents and Materials

Aseptically prepared
- LDF primary medium
- Holtfreter's buffer
- Dilute bleach, 0.1% in UPW
- Pronase E solution
- Feeder layers of embryonic fibroblasts

Protocol

Grow primary cultures derived from zebrafish blastula-stage and early gastrula-stage embryos on feeder layers of embryonic fibroblasts.

(a) Harvest embryos at the mid-blastula or early-gastrula stage and rinse several times with clean water.

(b) After rinsing, transfer the embryos into 60-mm petri dishes (50–100 embryos/dish), take to a laminar flow hood, and maintain under aseptic conditions.

(c) Soak the embryos for 2 min in dilute bleach and rinse several times in sterile Holtfreter's buffer.

(d) Dechorionate the embryos by incubating them in 2 ml of Pronase E solution for ~10 min, and then gently swirl the embryos in the petri dish to separate them from the partially digested chorion.

(e) Tilt the dish to collect the embryos on one side and gently remove 1.5 ml of the Pronase E solution with a Pasteur pipette. To prevent the dechorionated embryos from adhering to the dish and rupturing, keep the dish tilted so that the embryos remain suspended in the remaining Pronase solution.

(f) Gently rinse the embryos by adding 2 ml of Holtfreter's buffer and swirling gently.

(g) Tilt the dish and remove most of the Holtfreter's buffer, leaving the embryos suspended in ~0.5 ml of liquid.

(h) Repeat the rinse procedure twice more.

(i) After the final rinse, leave the embryos suspended in 0.5 ml of Holtfreter's buffer and add 2 ml of trypsin solution.

(j) Incubate the embryos in the trypsin for 1 min and dissociate the cells by gently pipetting 3–4 times.

(k) Immediately transfer the cell suspension into a sterile polypropylene centrifuge tube and add 200 μl of FBS to stop the trypsinization.

(l) Collect the cells by centrifugation (500 g, 5 min) and resuspend the pellet in LDF primary medium (without FBS or trout serum).

(m) Seed the cells at 1 × 10^4 cells/cm^2 onto feeder layers of growth-arrested embryonic fibroblasts, contained in multiwell dishes or flasks.

(n) Allow the cells to attach to the feeder layers (~15 min) before adding FBS and trout serum. Human recombinant leukemia inhibitory factor (10 ng/ml) is used in the medium in preference to BRL-conditioned medium [Sun et al., 1995a,b].

5.1.2. Cell Lines

Protocol 5.1.2.

Reagents and Materials
Aseptically prepared
• LDF maintenance medium
• ZEM-2 cells (or equivalent)

Protocol
(a) Grow cultures derived from early zebrafish embryos, such as ZEM-2, in LDF maintenance medium to ~70% confluency.

(b) Incubate at 26°C in ambient air.

(c) Change medium approximately every 5 days.

5.2. Calcium Phosphate Mediated Transfection

The procedure for transfection of zebrafish embryo cells is similar to methods employed with mammalian cells, except for minor modifications [Graham and van der Ebb, 1973; Wigler et al., 1980; Corsaro and Pearson, 1981; Helmrich et al., 1988].

Protocol 5.2

Reagents and Materials
- Plasmid DNA, circular, purified by cesium chloride centrifugation [Maniatis et al., 1982]
- Calcium chloride solution 0.25 M
- HEPES-buffered saline
- Tuberculin syringe with 25G needle
- Culture flask, 25 cm^2, of cells to be transfected. The flask should contain approximately 10^6 cells and have had the medium changed 4 h previously.

Protocol
(a) Suspend 25 μg of plasmid DNA in 0.5 ml 0.25 M calcium chloride solution.

(b) Add the suspension dropwise with constant mixing to a tube containing 0.5 ml HEPES-buffered saline. A DNA-calcium phosphate precipitate will form immediately.

(c) Incubate the mixture for 30 min at room temperature, then shear twice by passing it through a 25G needle attached to a tuberculin syringe.

(d) Add the precipitate-containing suspension to the 25-cm^2 flask containing the target cells. The DNA-calcium phosphate precipitate should be visible and evenly distributed on the surface of the cells.

(e) Incubate the cells in the presence of the DNA-calcium phosphate precipitate for a minimum of 6 h. The calcium phosphate solution is toxic when left on many mammalian cell lines for longer periods of time, especially in serum-free medium; however, in general, fish cells are able to tolerate longer periods of exposure. When transfecting ZEM-2 cells, the calcium phosphate solution can remain on the cells up to 12 h without a toxic effect.

(f) Change the medium to remove the precipitate and incubate the cells at 26°C.

Note: Optimal transfection of a larger population of cells (5 × 10^6) contained in a 75-cm^2 flask is achieved using 35–50 μg of plasmid DNA.

5.3. Lipofection (Liposome-Mediated Transfection)

Protocol 5.3

Reagents and Materials

Aseptically prepared

- Polypropylene tubes, 12 × 75 mm
- Plasmid DNA, circular, purified, 20 μg per transfection
- Lipofectin reagent, 50 μl per transfection
- Culture flask, 25 cm², containing approximately 10⁶ target cells
- LDF medium without serum or embryo extract
- LDF primary or maintenance medium
- UPW, 1 ml

Protocol

(a) Suspend 20 μg of plasmid DNA in 300 μl of UPW in the polypropylene tube.

(b) In a separate tube, add 50 μl of Lipofectin reagent (GIBCO BRL) to 200 μl of UPW.

(c) Combine the contents of each tube and mix gently.

(d) Incubate the DNA-lipofectin solution at room temperature for 15 min.

(e) Add the DNA-lipofectin solution to a 25-cm² flask containing approximately 10⁶ cells in fresh LDF medium without serum or embryo extract.

(f) After 12–24 h, replace the medium in the flask with fresh LDF medium containing serum and other supplements.

(g) Incubate the cells at 26°C.

5.4. Histochemical Detection of Cells Expressing the *lacZ* Reporter

Transient expression of plasmid DNA can be detected 2–4 days after transfection, depending on the cell line. For ZEM-2 cells, the highest level of expression occurs 2 days after transfection [Sharps et al., 1992]. A common reporter gene used to evaluate transient expression in zebrafish embryo cell cultures is *lacZ*, which can be detected using a spectrophotometric [Driever et al., 1993] or a histochemical [Sharps et al., 1992] assay. Results with the two assay methods usually correlate well; however, the histochemical assay is generally less sensitive, since it requires sufficient enzyme activity to generate a microscopically visible signal. An advantage of this method is that it can give an indication of the level of expression in individual cells and it can be used to identify particular plasmid constructs that will be suitable for the production of transgenic fish in which individual tissues will be examined histochemically. The following is a procedure to assay cells histochemically for *lacZ* expression [Sanes et al., 1986].

Protocol 5.4

Reagents and Materials
- PBSA
- Fixative solution (see Section 4.4)
- Staining buffer
- X-Gal stock solution

Protocol
(a) Rinse cultures 2× with 3 ml of PBSA and fix for 5 min in fixative solution at 4°C.
(b) After fixation, rinse the cells twice with PBSA and incubate for 16 h in staining buffer (Section 4.4) containing 1 mg/ml X-Gal (Section 4.4). Add a sufficient volume of the staining buffer to cover the culture surface and then add the X-Gal stock solution (20 μl X-Gal stock solution per milliliter of staining buffer).
(c) Identify the blue-colored *lacZ*-positive cells by examining the cultures by phase contrast microscopy.

Depending on the promoter and particular cell lines, fish cell transfection frequencies are, typically, 10^{-3} to 10^{-4} [Sharps et al., 1992].

5.5. Isolation of Stably Transfected Colonies

The bacterial selectable marker gene aminoglycoside phospho-transferase (*neo*), which confers resistance to the antibiotic neomycin and its analogue G418 (Geneticin), has been used to isolate colonies of stably transfected ZEM-2 cells [Collodi et al., 1992a]. The selection method used with the zebrafish cells has been adapted from procedures developed for use with mouse embryo cells [Shirahata et al., 1990]. This selection method can be used to isolate stable transfectants expressing a nonselectable gene in addition to *neo* by introducing *neo* in the same plasmid as the nonselectable gene, or by co-transfecting cells with two plasmids, each containing one of the two genes. For co-transfections, a 10-fold higher concentration of the plasmid containing the nonselectable gene is added compared to the *neo*-containing plasmid.

Protocol 5.5

Reagents and Materials

Aseptically prepared
- Tissue culture dishes, 100 mm diameter, 5
- Cloning rings
- Pasteur pipettes
- LDF maintenance medium
- LDF selection medium
- Trypsin, crude 0.2%

Protocol

(a) Transfect ZEM-2 cells grown in 75-cm^2 flasks (10^7 cells/flask) as described above.

(b) Two days after transfection, split the cells contained in each flask into five 100 mm diameter tissue culture dishes in LDF maintenance medium.

(c) Five days later remove the medium and replace it with fresh LDF selection medium.

(d) Non-transfected cells will begin to die 3–4 days after G418 is added. Change the medium every 3 days to remove dead cells and replace it with fresh LDF selection medium. After 7–10 days, most of the dead cells will be removed and the medium can be changed every 5 days.

(e) Approximately 1 week after the addition of G418, small colonies of resistant cells will be visible under the microscope. Allow the colonies to grow for an additional 2–3 weeks before transferring them from the dish.

(f) Isolate individual colonies using sterile glass cloning rings:

    (i) Remove the medium from the dish and rinse with PBSA

    (ii) Lightly coat one edge of the cloning ring with sterile silicon grease by dipping the end of the ring in the grease before putting it on the dish. The silicon grease will form a tight seal between the ring and culture dish.

    (iii) Gently remove any medium trapped inside the ring with a sterile Pasteur pipette and add 2–3 drops of trypsin solution to the ring.

    (iv) After a 30-s incubation in the trypsin, wash the cells contained in the ring off the dish by gently pipetting the trypsin solution up and down 2 or 3 times.

    (v) Transfer the trypsin solution with the suspended cells to a single well of a 24-well dish containing 200 μl of LDF without trout serum or FBS. Harvest the cells remaining on the original culture dish by adding 2–3 drops of LDF to the cloning ring and rinsing with a Pasteur pipette. Add the rinse to the contents of the 24-well plate and repeat the rinsing procedure.

(vi) Harvest other colonies in the same way.

(vii) After transferring the cells, let the 24-well dish sit undisturbed for 5 min to enable the cells to stick to the culture surface.

(viii) When the cells have attached to the plastic, add FBS (1%) and trout serum (0.5%) to the medium to inhibit the action of the trypsin.

(ix) After 1 h the cells will be tightly attached to the plate. Gently remove all of the medium in the well and replace it with fresh LDF maintenance medium.

(g) Twenty-four hours later change the medium again to remove residual trypsin and add fresh LDF selection medium.

(h) Change the medium every 5 days. When the culture becomes confluent, trypsinize the cells and transfer them to a single well on a 12-well plate. Gradually expand the culture to a 25-cm² flask and freeze a portion of the cells in liquid nitrogen at each passage.

Using the calcium phosphate method described above, stable transfection frequencies for ZEM-2 cells range from 1 to 5 cells per 10^5 transfected cells depending on the promoter and plasmid construct used. Transient expression frequencies are usually 10 to 100 times higher. Transfer of DNA into ZEM-2 cells by lipofection resulted in approximately the same transfection frequencies as achieved by the calcium phosphate method. Transfection frequencies of many mammalian cell lines are enhanced when a glycerol shock is included in the protocol, and this treatment has been reported to result in an increased uptake of plasmid DNA into ZF4 zebrafish embryo cells [Driever and Rangini, 1993]. However, the inclusion of a glycerol shock treatment during the transfection of ZEM-2 is toxic to the cells.

6. USE OF CULTURED EMBRYO CELLS FOR THE PRODUCTION OF CHIMERIC FISH

The transparency and accessibility of the zebrafish embryo make it suitable for experimental approaches involving cell transplantation. Embryo cells that have been genetically manipulated in culture by transfection or retroviral infection can be reintroduced into host embryos to produce chimeric fish [Collodi et al., 1992b; Sun et al., 1995b]. The full potential of this experimental approach will be realized when pluripotent embryonic stem (ES) or primordial germ cell lines are available from zebrafish for the production of germline chimeras. In mammals, ES cell cultures have been employed for efficient genome manipulation resulting in the overexpression or inactivation of particular genes [Cappechi, 1989; Gossler et al., 1986]. Zebrafish embryo cells obtained from established lines and long-term primary cultures are able to participate in normal development when introduced into host embryos by microinjection, but germline chimeras have only been produced with cells that were maintained for a very short period in primary culture [Sun et al., 1995b; Speksnijder et al., 1997; Gaiano et al., 1995]. The ability of donor cells to participate in normal host development has been evaluated by pigment formation [Gaiano et al., 1995; Lin et al., 1992], confocal microscopy [Speksnijder et al., 1997], and PCR analysis [Sun et al., 1995b]. To exploit this experimental approach involving the use of chimeras for the production of transgenic zebrafish, further work is needed to develop conditions that will maintain the pluripotency of the embryo cells during extended periods of culture.

REFERENCES

Amsterdam A, Lin S, Hopkins N (1995): The *Aequorea victoria* green fluorescent protein can be used as a reporter in live zebrafish embryos. Dev Biol 171: 123–129.

Amsterdam A, Lin S, Moss LG, Hopkins N (1996): Requirements for green fluorescent protein detection in transgenic zebrafish embryos. Gene 173: 99–103.

Bayer TA, Campos-Ortega JA (1992): A transgene containing *lacZ* is expressed in primary sensory neurons in zebrafish. Development 115: 421–426.

Bradford CS, Sun L, Barnes DW (1994a): Basic FGF stimulates proliferation and suppresses melanogenesis in cell cultures derived from early zebrafish embryos. Mol Mar Biol Biotech 3: 78–86.

Bradford CS, Sun L, Collodi P, Barnes DW (1994b): Cell cultures from zebrafish embryos and adult tissues. J Tissue Cult Methods 16: 99–107.

Bradford MM (1976) A rapid and sensitive method for the quantitation of microgram quantitites of protein utilizing the principle of protein-dye binding. Anal Biochem 72: 248–254.

Burns JC, Friedmann T, Driever W, Burrascano M, Yee J-K (1993): Vesicular stomatitis virus G glycoprotein pseudotyped retroviral vectors: concentration to very high titer and efficient gene transfer into mammalian and nonmammalian cells. Proc Natl Acad Sci USA 90: 8033–8037.

Caldovic L, Hackett PB (1995): Development of position-independent expression vectors and their transfer into transgenic fish. Mol Mar Biol Biotech 4: 51–61.

Capecchi MR (1989): Altering the genome by homologous recombination. Science 244: 1288–1292.

Collodi P, Barnes DW (1990): Mitogenic activity from trout embryos. Proc Natl Acad Sci USA 87: 3498–3502.

Collodi P, Kamei Y, Ernst T, Miranda C, Buhler DR, Barnes DW (1992a): Culture of cells from zebrafish (*Brachydanio rerio*) embryo and adult tissues. Cell Biol Toxicol 8: 43–61.

Collodi P, Kamei Y, Sharps A, Weber D, Barnes DW (1992b): Fish embryo cell cultures for derivation of stem cells and transgenic chimeras. Mol Mar Biol Biotech 1: 257–265.

Corsaro CM, Pearson ML (1981): Enhancing the efficiency of DNA-mediated gene transfer in mammalian cells. Somat Cell Genet 7: 603–616.

Culp P, Nusslein-Volhard C, Hopkins N (1991): High frequency germ-line transmission of plasmid DNA sequences injected into zebrafish eggs. Proc Natl Acad Sci USA 88: 7953–7957.

Culp P, Farrell M, Gaiano N, Lin S, Hopkins N (1995): Assessing the use of gene traps in zebrafish embryos and cultured cells (abstract). In: (Driever W, Eisen J, Grunwald D, Kimmel C, eds.) "Zebrafish Development and Genetics". Cold Spring Harbor Laboratory Press, Cold Spring Harbor, New York p. 110.

Driever W, Rangini Z (1993): Characterization of a cell line derived from zebrafish (*Brachydanio rerio*). In Vitro Cell Dev Biol Anim 29A: 749–754.

Driever W, Stemple D, Schier A, Solnica-Krezel L (1994): Zebrafish: genetic tools for studying vertebrate development. Trends Genet 10: 152–159.

Evans MJ, Kaufman MH (1981): Establishment in culture of pluripotential cells from mouse embryos. Nature 292: 154–156.

Fu L, Aoki T (1991): Expression of CAT gene directed by the 5'-upstream region from yellowtail (α-globin gene in tissue cultured fish cells. Nippon Suisan Gakkaishi 57: 1689–1696.

Furutani-Seiki M, Grandel H, Beuchle D, Herr U, Vogelsang E, Nusslein-Volhard C (1995): Towards generating an embryonic stem (ES) cell line from zebrafish (abstract). In: (Driever W, Eisen J, Grunwald D, Kimmel C, eds.) "Zebrafish Development and Genetics". Cold Spring Harbor Laboratory, p. 113.

Gaiano N, Chen J, Long W, Amsterdam A, Hopkins N (1995): Blastomeres cultured for short periods of time can be used to generate germ-line chimeric zebrafish (abstract). In: (Driever W, Eisen J, Grunwald D, Kimmel C, eds.) "Zebrafish Development and Genetics". Cold Spring Harbor Laboratory, p. 112.

Gaiano N, Allende M, Amsterdam A, Kawakami K, Hopkins N (1996a): Highly efficient germ-line transmission of proviral insertions in zebrafish. Proc Natl Acad Sci USA 93: 7777–7782.

Gaiano N, Amsterdam A, Kawakami K, Allende M, Becker T, Hopkins N (1996b): Insertional mutagenesis and rapid cloning of essential genes in zebrafish. Nature 383: 829–832.

Ghosh C, Collodi P (1994): Culture of cells from zebrafish (*Brachydanio rerio*) blastula-stage embryos. Cytotechnology 14: 21–26.

Ghosh C, Liu Y, Ma C, Collodi P (1997): Cell cultures derived from early zebrafish embryos differentiate in vitro into neurons and astrocytes. Cytotechnology 23: 221–230.

Gibbs PDL, Peek A, Thorgaard G (1994): An in vivo screen for the luciferase transgene in zebrafish. Mol Mar Biol Biotech 3: 307–316.

Gossler A, Doetschman T, Korn R, Serfling E, Kemler R (1986): Transgenesis by means of blastocyst-derived embryonic stem cell lines. Proc Natl Acad Sci USA 83: 9065–9069.

Graham FL, van der Ebb AJ (1973): A new technique for the assay of infectivity of human adenovirus 5 DNA. Virology 52: 456–467.

Helmrich A, Bailey GS, Barnes DW (1988): Transfection of cultured fish cells with exogenous DNA. Cytotechnology 1: 215–221.

Ho RK, Kimmel CB (1993): Commitment of cell fate in the early zebrafish embryo. Science 26: 109–111.

Hopkins N (1993): High titers of retrovirus (vesicular stomatitis virus) pseudo-types, at last. Proc Natl Acad Sci USA 90: 8759–8760.

Ivics Z, Izsvak Z, Hackett PB (1993): Enhanced incorporation of transgenic DNA into zebrafish chromosomes by a retroviral integration protein. Mol Mar Biol Biotech 2: 162–173.

Ivics Z, Izsvak Z, Minter A, Hackett PB (1996): Identification of functional domains and evolution of Tc1-like transposable elements. Proc Natl Acad Sci USA 93: 5008–5012.

Izsvak, Z, Ivics, Z, Hackett PB (1995): Characterization of a Tc1-like transposable element in zebrafish (*Danio rerio*). Mol Gen Genet 247: 312–322.

Izsvak Z, Ivics Z, Garcia-Estefania D, Fahrenkrug SC, Hackett PB (1996): DANA elements: a family of composite, tRNA-derived short interspersed DNA elements associated with mutational activities in zebrafish (*Danio rerio*). Proc Natl Acad Sci USA 93: 1077–1081.

Kane DA, Warga RM, Kimmel CB (1992): Mitotic domains in the early embryo of the zebrafish. Nature 360: 735–737.

Lin S, Long W, Chen J, Hopkins N (1992): Germ line chimeras in zebrafish by cell transplants from genetically pigmented to albino embryos. Proc Natl Acad Sci USA 89: 4519–4523.

Lin S, Gaiano N, Culp P, Burns JC, Friedmann T, Yee J-K, Hopkins N (1994a): Integration and germ-line transmission of a pseudotyped retroviral vector in zebrafish. Science 265: 666–669.

Lin S, Yang S, Hopkins N (1994b): *lacZ* Expression in germline transgenic zebrafish can be detected in living embryos. Dev Biol 161: 77–83.

Maniatis T, Fritsch EF, Sambrook J (1982): "Molecular Cloning, a Laboratory Manual." Cold Spring Harbor Laboratory, NY: Cold Spring Harbor Press.

Martin GR (1981): Isolation of a pluripotent cell line from early mouse embryos cultured in medium conditioned by teratocarcinoma stem cells. Proc Natl Acad Sci USA 78: 7634–7640.

Moav B, Liu Z, Caldovic LD, Gross ML, Faras AJ, Hackett PB (1993): Regulation of expression of transgenes in developing fish. Transgenic Res 2: 153–161.

Moss JB, Price AL, Raz E, Driever W, Rosenthal N (1996): Green fluorescent protein marks skeletal muscle in murine cell lines and zebrafish. Gene 173: 89–98.

Mullins MC, Nusslein-Volhard C (1993): Mutational approaches to studying embryonic pattern formation in the zebrafish. Curr Opin Genet Dev 3: 648–654.

Nusslein-Volhard C (1994): Of flies and fishes. Science 266: 572–574.

Peppelenbosch MP, Tertoolen LGJ, DeLaat SW, Zivkovic D (1995): Ionic responses to epidermal growth factor in zebrafish cells. Exp Cell Res 218: 183–188.

Powers DA (1989): Fish as model systems. Science 246: 352–357.

Rossant J, Hopkins N (1992): Of fin and fur: mutational analysis of vertebrate embryonic development. Genes Dev 6: 1–13.

Sanes JR, Rubenstein JLR, Nicolas J-F (1986): Use of a recombinant retrovirus to study post-implantation cell lineage in mouse embryos. EMBO 5: 3133–3142.

Sharps A, Nishiyama K, Collodi P, Barnes DW (1992): Comparison of activities of mammalian viral promoters directing gene expression in vitro in zebrafish and other fish cell lines. Mol Mar Biol Biotech 1: 426–431.

Shirahata S, Rawson C, Loo D, Chang Y-J, Barnes DW (1990): *Ras* and *Neu* oncogenes reverse serum inhibition and epidermal growth factor dependence of serum-free mouse embryo cells. J Cell Physiol 144: 69–76.

Speksnijder JE, Hage WJ, Lanser PH, Collodi P, Zivkovic D (1997): In vivo assay for the developmental competence of embryo-derived zebrafish cell lines. Mol Mar Biol Biotech 6: 21–32.

Streisinger G, Walker C, Dower N, Knauber D, Singer F (1981): Production of homozygous diploid zebrafish (*Brachidanio rerio*). Nature 291: 293–296.

Stuart GW, McMurray JV, Westerfield M (1988): Replication, integration and stable germ-line transmission of foreign sequences injected into early zebrafish embryos. Development 103: 403–412.

Stuart GW, Vielkind JR, McMurray JV, Westerfield M (1990): Stable lines of transgenic zebrafish exhibit reproducible patterns of transgene expression. Development 109: 577–584.

Sun L, Bradford CS, Barnes DW (1995a): Feeder cell cultures for zebrafish embryonal cells in vitro. Mol Mar Biol Biotech 4: 43–50.

Sun L, Bradford CS, Ghosh, C, Collodi P, Barnes DW (1995b): ES-like cell cultures derived from early zebrafish embryos. Mol Mar Biol Biotech 4: 193–199.

Takagi S, Sasado T, Tamiya G, Ozato K, Wakamatsu Y, Takeshita A, Kimura M (1994): An efficient expression vector for transgenic medaka construction. Mol Mar Biol Biotech 3: 192–199.

Westerfield M (1993): "The Zebrafish Book: A guide for the laboratory use of zebrafish (*Brachydanio rerio*)." Eugene, OR: University of Oregon Press.

Wigler M, Perucho M, Kurtz D, Dana S, Pellicer A, Axel R, Silverstein S (1980): Transformation of mammalian cells with an amplifiable dominant acting gene. Proc Natl Acad Sci USA 77: 3567–3570.

Wu L, Schneider J, Ma C, Collodi P (1997): Transforming growth factor-β and retinoic acid regulate fibronectin synthesis in a zebrafish embryo cell line. Exp Cell Research (submitted).

APPENDIX: LIST OF SUPPLIERS

Item	Supplier
Ampicillin	Sigma
BRL cell line	American Type Culture Collection
Trypsin	GIBCO BRL
Fibroblast growth factor, bovine	R&D Systems
FBS	Harlan Bioproducts
G418 (Geneticin)	GIBCO BRL
Glass cloning rings	Bellco
Insulin	Upstate Biotechnology
Lipofectin reagent	GIBCO BRL
Media: Ham's F12, DMEM, L-15	GIBCO BRL
Mitomycin C	Sigma
Penicillin	Sigma
Plasmids	GIBCO BRL
Pronase E	Sigma
Streptomycin	Sigma
Tissue culture supplies (Falcon) (plastic pipettes, flasks, filters, dishes)	Fisher
Tissuemizer homogenizer	Tekmar
Trout serum	East Coast Biologics
Trypsin	Sigma
X-Gal	GIBCO BRL
ZEM-2 cell line	Available from author

5

Transfection of Rat Adipose Cells by Electroporation

Michael J. Quon

Hypertension-Endocrine Branch, National Heart, Lung, and Blood Institute, National Institutes of Health, 10 Center Drive, Bethesda, Maryland 20892-1754

DNA Transfer to Cultured Cells, Edited by Katya Ravid and R. Ian Freshney.
ISBN 0-471-16572-7 © 1998 Wiley-Liss, Inc.

I. INTRODUCTION

Adipose cells (adipocytes) are terminally differentiated, specialized cells whose primary physiological role has classically been described as an energy reservoir for the body. Adipocytes are a storage depot for triglycerides in times of energy excess and a source of energy in the form of free fatty acids released by lipolysis during times of energy need. Recently, the important role of adipose cells as active regulators of carbohydrate and lipid metabolism has received increased attention [Flier, 1995; Reaven, 1995]. It is likely that specific abnormalities in adipose tissue can contribute directly to the pathogenesis of common diseases such as diabetes, hypertension, and obesity [Flier, 1995; Hotamisligil et al., 1993, 1995; Hotamisligil and Spiegelman, 1994; Reaven, 1995; Walston et al., 1995; Zhang et al., 1994].

One of the best-characterized systems for understanding hormonal regulation of metabolism is the isolated rat adipose cell preparation first described by Rodbell [Rodbell, 1964]. However, technical difficulties encountered with manipulating and transfecting isolated adipose cells led to the development of alternative model systems such as the 3T3-L1 or 3T3-F442A tissue culture cell lines. These cells are derivatives of primitive mouse mesodermal cells that can differentiate into adipocyte-like phenotypes under the appropriate conditions [Spiegelman et al., 1993]. These tissue culture cells have been extremely useful for understanding the regulation and control of adipocyte differentiation by transcription factors such as PPARγ2 and C/EBP [MacDougald and Lane, 1995; Tontonoz et al., 1994]. However, these cell lines do not always differentiate completely and uniformly. Furthermore, they do not display the full repertoire of genes expressed in primary adipose cells. For example, ob gene mRNA levels in differentiated 3T3-L1 cells are only ~1% of the levels seen in freshly isolated rat adipose cells. Finally, these tissue culture cells are much less responsive than isolated adipose cells with respect to the effects of insulin and other hormones on glucose transport and metabolism. Therefore, it was of interest to develop an effective method for transfection of adipose cells in primary culture in order to apply modern methods in molecular biology to the study of a *bona fide* insulin target cell.

The electroporation protocol for adipose cells in primary culture described in this chapter provides a reliable means to express, effectively, recombinant genes in a physiologically relevant insulin target cell. This method has contributed to the understanding of insulin signal transduction pathways related to glucose transport and metabolism [Quon et al., 1993; 1994a,b; 1995]. In addition, this method is useful for studying the regulation of genes that are specifically expressed in adipose cells [He et al., 1995; Ing et al., 1996].

2. PREPARATION OF REAGENTS

Sterile

DMEM-A

Dulbecco's modified Eagle's Medium, pH 7.4 containing 25 mM glucose, 2 mM glutamine, 200 nM (R)-N^6-(1-methyl-2-phenylethyl)adenosine, 100 μg/ml gentamicin, and 25 mM HEPES. Add glutamine, gentamicin, PIA, and HEPES, to DMEM, warm to 37°C and adjust the pH to 7.4. Store at 4°C for up to 3 weeks.

DMEM-B

To make 100 ml of DMEM-B, add 7 g of bovine serum albumin to each 100 ml of DMEM-A, warm to 37°C, and adjust the pH to 7.4. Filter sterilize and store at 4°C for up to 3 weeks.

KRBH buffer

Krebs-Ringer medium, pH 7.4, containing 10 mM $NaHCO_3$, 30 mM HEPES (Sigma), 200 nM adenosine (Boehringer-Mannheim), and 1% (w/v) bovine serum albumin (Intergen). To make 1 liter of KRBH buffer, add 7 g of NaCl, 0.55 g of KH_2PO_4, 0.25 g of $MgSO_4 \cdot 7H_2O$, 0.84 g of $NaHCO_3$, 0.11 g of $CaCl_2$, 7.15 g of HEPES, 10 g of bovine serum albumin, 1 ml of 200 μM adenosine, and distilled water to a volume of 1 liter. Warm to 37°C and adjust the pH to 7.4

KRBH-A buffer

Same as KRBH except with 5% bovine serum albumin

Collagenase, 20 mg/ml

Dissolve 40 mg in 2 ml of KRBH buffer and sterilize by passing through a 0.22-μM filter

Nonsterile

Extraction solution M

Mix 800 ml of 2-propanol, 200 ml of n-heptane, and 20 ml of 1 N sulfuric acid.

KCN solution

Mix 5.2 mg of KCN, 850 μl of KRBH-A buffer, and 150 μl of 0.1 N HCl.

3. PREPARATION OF ISOLATED RAT ADIPOSE CELLS

3.1. Dissection of Epididymal Fat Pads

Protocol 3.1

Reagents and Materials

Aseptically prepared

- Scissors: two pairs of sharp dissecting scissors, 10 cm (4 in)
- Perry forceps, 12.5 cm (5 in)

Nonsterile

- Obtain male Sprague-Dawley rats (145–170 g, CD strain, Charles River Breeding Laboratories) 1 week prior to use. House the animals in cages with pine shavings in an animal care facility with *ad libitum* food and water until the morning of the experiment.
- Plastic box perfused with a gas mixture of 70% CO_2, 30% O_2
- Guillotine
- 70% ethanol

Protocol

(a) Anesthetize rats in a plastic box with a gas mixture of 70% CO_2, 30% O_2.

(b) Decapitate rats using a guillotine and exsanguinate.

(c) Soak the bodies briefly in 70% ethanol.

(d) Remove the epididymal fat pads maintaining the highest level of sterility possible.

   (i)   Cut through the skin on the lower abdomen with one pair of scissors to expose the peritoneum.

   (ii) Using a second pair of scissors, open the peritoneum and pull up the testes with a pair of forceps.

   (iii) Trim the fat pads from the epididymi, taking care to leave the blood vessels behind.

(e) Transport the tissue to the culture laboratory.

3.2. Collagenase Digestion and Washing of Adipose Cells

To obtain isolated adipose cells we use a modification of the procedure originally described by Rodbell [Rodbell, 1964]. The procedure described here is optimized to obtain approximately 4 ml of packed adipose cells from the epididymal fat pads of four rats and can be scaled up or down as needed.

Protocol 3.2

Reagents and Materials

Aseptically prepared
- DMEM-A medium at 37°C, 100 ml
- KRBH buffer at 37°C, 100 ml
- Collagenase, 20 mg/ml
- Sharp dissecting scissors
- Nylon mesh filter, 250 μm, in holder or filter funnel
- Low-density polypropylene vial, 30 ml
- Conical centrifuge tube, 50 ml (Corning)

Nonsterile
- Tabletop centrifuge

Protocol

(a) Add 4 g of fat pads (approximately equivalent to 8 fat pads) to a 30-ml low-density polypropylene vial containing 4 ml of KRBH buffer at 37°C.

(b) Mince the fat pads into pieces approximately 2 mm in diameter with scissors.

(c) Add 1 ml of collagenase solution to the vial containing the minced fat pads and incubate in a shaking water bath at 37°C for approximately 1 h until the cell mixture has a creamy consistency.

(d) After collagenase digestion, add 10 ml of KRBH buffer at 37°C to the vial.

(e) Mix the cells in the vial by swirling, and gently pass through a 250-μm nylon mesh filter into a 50-ml conical tube.

(f) Wash the cells by adding 30 ml of KRBH buffer at 37°C to the tube and centrifuging briefly at $200 \times g$ in a tabletop centrifuge and remove the infranatant with a pipette. Note that the adipose cells will be floating on top of the aqueous buffer.

(g) Repeat the washing of cells by adding 40 ml of KRBH buffer, centrifuging, and removing the infranatant two additional times.

(h) Then wash the cells twice with 40 ml of DMEM-A medium at 37°C.

(i) After the final wash, resuspend the cells in DMEM-A medium at a cytocrit of approximately 40% and transfer to a 30-ml low-density polypropylene tube containing a magnetic stir bar.

(j) Take a 0.5-ml aliquot of cells and perform a glucose uptake assay to check cell viability (see Section 6).

4. ELECTROPORATION

4.1. Preparation of Plasmid DNA

Obtain milligram quantities of supercoiled plasmid DNA using the Wizard Megaprep kit. Determine the concentration of plasmid

DNA by comparison with known DNA markers, using ethidium bromide staining of restriction-digested plasmid run on an agarose gel. For overexpression of recombinant genes, we typically use the expression vector pCIS2 that contains a CMV promoter/enhancer with a generic intron upstream of the multiple cloning site [Choi et al., 1991]. We have previously shown that this expression vector works well in adipose cells and results in 20- to 40-fold overexpression of various recombinant genes [Ing et al., 1996; Quon et al., 1993; 1994a,b; 1995].

4.2. Preparation of Electroporation Cuvettes

Protocol 4.2

Reagents and Materials

Aseptically prepared
- PBS: Dulbecco's PBS with Ca^{2+} and Mg^{2+} (BioFluids)
- Cuvette with 0.4-cm internal diameter (BioRad)
- Plasmid DNA: See above, Section 4.1
- Low-density polypropylene flat-bottomed vial, 30 ml
- Micropipette with wide-bore pipette tip, 200 μl

Protocol
(a) Place 0.2 ml of PBS in a 0.4-cm gap cuvette.
(b) Add up to 9 μg of plasmid DNA. Adding more than 9 μg of DNA per cuvette significantly affects cell viability.
(c) Mix adipose cells in a 30-ml tube using a magnetic stirring bar.
(d) Transfer 0.2 ml of the 40% cytocrit preparation to each cuvette using a 200-μl wide-bore pipette tip.
(e) Mix the cells with the DNA by pipetting up and down in the cuvette.

4.3. Electroporation

Protocol 4.3

Reagents and Materials
- Gene Pulser II Transfection Apparatus (BioRad)

Protocol
(a) Ten minutes after adding cells to the electroporation cuvettes, transfect by administering three shocks on transfection apparatus at a voltage of 800 V and a capacitance of 25 μF.
(b) Reverse the polarity of electroporation and administer three more shocks. The time constant of electroporation is typically ~0.6 ms during the final shock.

Do not process more than approximately 70 cuvettes for each experiment, as the viability of the cells will be affected by the duration of the electroporation procedure. We estimate that our electroporation protocol results in the transient transfection of approximately 5% of the cells [Quon et al., 1994b].

5. PRIMARY CULTURE OF ADIPOSE CELLS

Protocol 5

Reagents and Materials
Aseptically prepared
- DMEM-B medium, 0.5 ml per cuvette
- Polypropylene tubes, 17 mm × 100 mm
- Wide-bore pipette tips, 200 μl
- Tissue culture petri dish, 60-mm diameter

Nonsterile
- Micropipette
- Humid incubator at 37°C with 5% CO_2

Protocol
(a) Ten minutes after electroporation, transfer the contents of the cuvettes to polypropylene tubes, using 200-μl wide-bore pipette tips. If cells from multiple cuvettes are to be pooled, the contents from up to 15 cuvettes can be placed into one 60-mm tissue culture dish (Becton Dickinson).
(b) Place the cells in a humid incubator at 37°C with 5% CO_2.
(c) Add 5 ml of DMEM-B medium to each tube or dish 1.5 h after electroporation.

After 16–24 h of incubation the cells are ready for further studies. At this point, we typically perform a glucose uptake assay on an aliquot of cells to check cell viability (see Section 6).

6. D-[U-^{14}C]GLUCOSE UPTAKE ASSAY

The assay described here measures D-[U-^{14}C]glucose incorporation into adipose cells in response to insulin. Since transport is the rate-limiting step for glucose incorporation under our conditions, this assay gives an estimate of glucose transport. Although this assay is not as sophisticated as the 3-*O*-methylglucose transport assay [Karnieli et al., 1981], it is a quick, easy, and effective method for assessing the ability of adipose cells to respond to the metabolic actions of insulin.

We usually perform this assay in triplicate with three determinations of basal glucose uptake and three determinations of insulin-stimulated glucose uptake for each group of cells.

Protocol 6

Reagents and Materials
- KRBH buffer, 10 ml
- KRBH buffer with or without 60 nM insulin, 200 μl per sample
- Polyethylene vials, 7 ml (Research Products)
- KRBH buffer containing D-[U-^{14}C]glucose (Dupont NEN): 12.5 μl of D-[U-^{14}C] glucose with a specific activity of 294 mCi/mmol (~11 GBq/mmol) diluted in 5 ml of KRBH buffer; 5 ml per sample
- Microcentrifuge tubes, 500 μl, 4 mm $\times$ 45 mm (Apple Scientific)
- Dinonylphthalate oil (ICN Biomedicals)
- Microcentrifuge
- Rotor, vertical
- Scintillation counter
- Shaking water bath at 37°C

Protocol
(a) Dilute an aliquot of cells with KRBH buffer to a cytocrit of 5%.
(b) Add 200 μl of KRBH buffer, with or without 60 nM insulin, to 7-ml polyethylene vials.
(c) Add 200-μl aliquots of the 5% cell suspension to the tubes.
(d) After incubation for 30 min at 37°C in a shaking water bath, add 100 μl of KRBH buffer containing D-[U-^{14}C]-glucose to each vial.
(e) Add 100 μl of dinonylphthalate oil to microcentrifuge tubes.
(f) After a 30-min incubation with D-[U-^{14}C]-glucose at 37°C, transfer 300-μl aliquots to microcentrifuge tubes containing dinonyl-phthalate oil.
(g) Rapidly separate cells from the aqueous buffer by centrifugation at 10,000 $\times$ g for 30 s. Cells will be in the top layer, separated from the aqueous buffer by the dinonylphthalate oil.
(h) Count cell-associated radioactivity in a liquid scintillation counter. The cell-associated radioactivity is a measure of glucose uptake by the cell.

To normalize the data for the number of cells in each sample, determine the lipid content of a 200-μl aliquot of cells from the original 5% suspension [Cushman and Salans, 1978] (see Section 8). For freshly isolated adipose cells we typically see a 10- to 15-fold increase in glucose uptake after maximal insulin stimulation. For adipose cells that have undergone electroporation and culture for 1 day, we typically see a 5-fold increase in glucose uptake with insulin stimulation. This assay tends to underestimate glucose transport. When we measured glucose transport in our cells using a 3-*O*-methylglucose transport assay, freshly isolated cells had a 20-fold increase in their response to insulin, while electroporated, cultured adipose cells had a 10-fold increase in their response to insulin [Quon et al., 1993].

7. DOUBLE-ANTIBODY ASSAY FOR CELL SURFACE EPITOPE-TAGGED GLUT4

To study GLUT4 translocation exclusively in the small fraction of cells that are transfected, we have constructed an expression vector for GLUT4 that has the influenza hemagglutinin epitope placed in the first exofacial loop of GLUT4 (GLUT4-HA) [Quon et al., 1994b]. We can generate insulin dose-response curves for the transfected cells using a double-antibody assay to quantitate the amount of GLUT4-HA at the cell surface in response to insulin.

Protocol 7

Reagents and Materials
Aseptically prepared
- GLUT4-HA (see Section 7 above)
- Empty expression vector pCIS2
- KRBH-A buffer at 37°C, 30 ml per tube
- KRBH-A buffer with 0.5 mg/ml bacitracin (to inhibit degradation of insulin at low concentrations), 5 ml per tube
- Insulin in KRBH-A buffer: serial dilutions of 2.4, 7.2, 30 nM, and 6 μM
- Tissue culture petri dishes, 60 mm
- Conical polystyrene centrifuge tubes, 15 ml (Corning)
- Polyethylene vials, 20 ml (Apple Scientific).

Nonsterile
- Shaking water bath at 37°C
- KCN solution, 80 mM (handle with care, highly toxic)
- Anti-HA1 mouse monoclonal antibody 12CA5 (Boehringer-Mannheim)
- [^{125}I]-sheep anti–mouse IgG, 370 kBq (10 μCi)/μg, final dilution 1 : 500 (Amersham)
- Dinonylphthalate oil.
- Polyethylene vials, 7 ml (Apple Scientific).
- Microcentrifuge tubes (4 mm × 45 mm)

- Microcentrifuge (Beckman, Heraeus)
- Gamma counter

Protocol

(a) Transfect one group of adipose cells, as described above, with 3 μg/cuvette of GLUT4-HA.

(b) Pool the cells from 20 cuvettes in 60-mm tissue culture dishes (10 cuvettes per dish), and culture as described above.

(c) In addition, transfect and culture another group of adipose cells with 3 μg/cuvette of the empty expression vector pCIS2 for use in determining nonspecific binding (10 cuvettes).

(d) Transfer cells, 24 h after electroporation, to 15-ml conical polystyrene tubes and wash with 13 ml of KRBH-A buffer at 37°C.

(e) Resuspend cells to a cytocrit of 15% using KRBH-A buffer with 0.5 mg/ml bacitracin.

(f) Place 2-ml aliquots of the cell suspension in five 20-ml polyethylene vials.

(g) Add 20 μl of serial dilutions of an insulin solution in KRBH-A buffer so that final insulin concentrations of 0, 0.024, 0.072, 0.3, and 60 nM are obtained.

(h) Incubate the cells in a water bath at 37°C for 30 min with gentle shaking.

(i) To prevent redistribution of GLUT4 after insulin stimulation, add 50 μl of KCN solution, to give a final concentration of 2.0 mM, and incubate the cells for an additional 5 min.

(j) Incubate the cells for 1 h at 24°C with the anti-HA1 mouse monoclonal antibody 12CA5 at a final dilution of 1:4000.

(k) Following incubation with 12CA5, wash the cells twice with 13 ml of KRBH-A buffer at 24°C.

(l) Resuspend the cells in a final volume of 0.7 ml KRBH-A buffer and transfer to 7-ml polyethylene vials.

(m) Remove a 200-μl aliquot from each vial for determination of lipid weight (see Section 8, below).

(n) Incubate with [^{125}I]-sheep anti–mouse IgG, final dilution 1:500, for 1 h at 24°C.

(o) Add 100 μl of dinonylphthalate oil to microcentrifuge tubes

(p) Transfer 300-μl aliquots of incubation mixture to microcentrifuge tubes containing 100 μl of dinonylphthalate oil.

(q) Separate the cells from the medium by centrifugation at 10,000 × g for 30 s. Note that cells will be in the top layer, separated from the aqueous media by a layer of dinonylphthalate oil.

(r) Count cell-associated radioactivity in a gamma counter.

Cells transfected with the empty expression vector pCIS2 are used to determine nonspecific binding of the antibodies. Typically, the nonspecific binding is ~30% of the total binding to cells transfected with GLUT4-HA and maximally stimulated with insulin [Quon et al., 1994b]. The lipid content from a 200-μl aliquot of cells is used to normalize the data for each sample.

8. DETERMINATION OF LIPID CONTENT OF ADIPOSE CELLS

This method has been described previously (Cushman and Salans, 1978).

Protocol 8

Reagents and Materials
Nonsterile
- Extraction solution M, 3 ml per sample tube (see Section 2, above)
- *n*-Heptane, 2 ml per sample
- Distilled water, 2 ml per sample
- Glass test tubes, 16 × 100 mm
- Preweighed glass test tubes, 12 × 75 mm
- Vortex mixer
- Balance, μg sensitivity

Protocol
(a) For each determination of lipid content, prepare a 16 × 100 mm glass test tube containing 2.7 ml of extraction solution M and place a rubber stopper on top to prevent evaporation.
(b) To each tube, add a 200-μl aliquot of the adipose cell suspension whose lipid content is being measured and shake gently.
(c) Incubate at room temperature for 1 h.
(d) Add 1.8 ml of *n*-heptane and 1 ml of distilled water to each tube.
(e) Mix by vortexing and then centrifuge for 5 min at 400 g.
(f) Place 1 ml of the organic phase (top layer) in an empty 12 × 75 mm glass test tube that has been previously weighed. Evaporate the heptane, leaving the extracted lipid in the test tube.
(g) Re-weigh the test tube to determine the amount of lipid contained in each aliquot of cells.

9. APPLICATIONS: INVESTIGATION OF INSULIN SIGNAL TRANSDUCTION PATHWAYS RELATED TO GLUCOSE TRANSPORT

Adipose cells are among the most responsive cell types with respect to insulin-stimulated glucose transport. The ability of insulin to recruit the insulin-responsive glucose transporter GLUT4 from an intracellular compartment to the cell surface accounts for the majority of the increase in glucose uptake seen with insulin stimulation. GLUT4 is expressed primarily in muscle and adipose tissue, and the ability of insulin to stimulate translocation of GLUT4 appears to be tissue specific. For example, insulin treatment does not generally result in a large effect on GLUT4 translocation in cell types such as CHO, COS, or NIH-3T3 cells that have been transfected with GLUT4, even if these cells are also overexpressing insulin receptors. Therefore, our adipose cell transfection system provides an opportunity to elucidate insulin signaling pathways related to glucose transport in a physiologically relevant cell. Since our electroporation protocol results in the transient transfection of only 5% of the cells, we cotransfect the cells with GLUT4-HA, an expression vector for GLUT4 with an epitope-tag from influenza hemagglutinin (HA) inserted in the first exofacial loop of GLUT4 [Quon et al., 1994b]. This allows us to study GLUT4 translocation exclusively in the small fraction of cells that are transfected. Using the double-antibody assay described above, we are able to quantitate the amount of GLUT4-HA on the surface of transfected cells in response to insulin. Thus, we can evaluate the role of various molecules in insulin-stimulated translocation of GLUT4 by overexpressing wild-type genes, inhibiting endogenous gene products with mutants that behave in a dominant negative manner, or using antisense ribozymes designed to interfere with production of endogenous gene products [Quon et al., 1994a,b; 1995].

We typically cotransfect adipose cells with 6 μg/cuvette of the expression vector containing the cDNA of interest, and 3 μg/cuvette of GLUT4-HA to increase the likelihood that cells expressing GLUT4-HA will also be expressing the genes of interest. Under these conditions, >95% of the cells expressing GLUT4-HA also express the gene of interest [Quon et al., 1995]. Cells from 20 to 30 cuvettes are pooled for each group of a typical insulin dose-response experiment. A group of cells transfected with the empty expression vector pCIS2 is used to estimate nonspecific binding of the antibodies.

9.1. Role of Insulin Receptor Tyrosine Kinase in Translocation of GLUT4

To study the role of the insulin receptor tyrosine kinase in GLUT4 translocation we overexpressed either wild-type human insulin receptors or receptors with an inactive kinase domain (substitution of Ile for Met at position 1153) in adipose cells along with GLUT4-HA; we then measured cell surface GLUT4-HA in response to insulin using the double-antibody assay described above (Fig. 5.1) [Quon et al., 1994b]. Control cells co-transfected with the empty expression vector pCIS2 and GLUT4-HA show a 4-fold increase in cell surface epitope-tagged GLUT4 in response to maximal insulin stimulation. The half-maximal insulin dose is approximately 50 pM. Cells co-transfected with the wild-type human insulin receptor (pCIS-hIR) and GLUT4-HA showed an elevation in the basal cell surface epitope-tagged GLUT4 to a level near that seen with maximal insulin stimulation. Cells co-transfected with the kinase inactive mutant insulin receptor (pCIS-Ile) and GLUT4-HA had an insulin dose-response similar

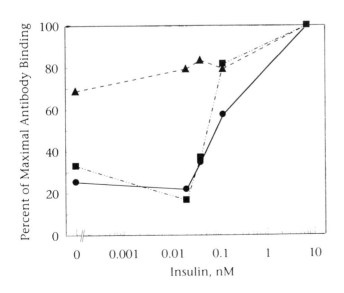

Fig. 5.1. Insulin dose response in co-transfected adipose cells (composite of two representative experiments). *Filled circle:* Cells co-transfected with pCIS2/GLUT4-HA (6 μg and 3 μg/cuvette). *Filled triangle:* pCIS-hIR/GLUT4-HA (6 μg and 3 μg/cuvette). *Filled square:* pCIS-Ile/GLUT4-HA (6 μg and 3 μg/cuvette). All cells were processed as described. In one experiment, insulin doses of 0, 0.02, 0.04, and 7 nM were used, while another experiment used insulin doses of 0, 0.04, 0.12, and 7 nM. Mean values are shown for the insulin doses of 0, 0.04, and 7 nM. Results obtained in cells transfected with pCIS2 alone represent nonspecific binding and were used to calculate the specific binding in the other samples. Results are expressed as the percent of maximal specific antibody binding. Absolute counts measured with maximal insulin stimulation were similar for all three conditions in which cells were transfected with pCIS-GLUT4-HA: 737, 644, and 810 cpm in cells co-transfected with pCIS-hIR, pCIS-Ile, or PCIS2 respectively.

to that of the control cells. These results suggest that normal insulin receptors possess intrinsic activity in the basal state. In the presence of a sufficient excess of receptors, the total basal activity induces near-maximal recruitment of GLUT4 even in the absence of insulin. The fact that cells co-transfected with the tyrosine-kinase deficient receptor do not have an elevation in basal cell surface GLUT4-HA suggests that the intrinsic basal activity of the normal receptor is associated with tyrosine kinase activity. Furthermore, the inability of the mutant to signal any detectable GLUT4-HA recruitment (in the presence or absence of insulin) strongly suggests a direct role for receptor tyrosine kinase activity in insulin-stimulated glucose transport in adipose cells.

9.2. Role of Phosphatidylinositol 3-Kinase in Translocation of GLUT4

Insulin action is initiated by the binding of insulin to its receptor, followed by activation of receptor tyrosine kinase activity and subsequent phosphorylation of cellular substrates such as IRS-1. We previously demonstrated a role for IRS-1 in insulin-stimulated translocation of GLUT4 in transfected adipose cells [Quon et al.,

Fig. 5.2. Recruitment of epitope-tagged GLUT4 to the cell surface of isolated rat adipose cells co-transfected with either SRα-Δp85/GLUT4-HA (*filled triangle*) or pCIS2/GLUT4-HA (*filled circle*). Results are the means ± SEM of four independent experiments. Data are expressed as a percentage of cell surface GLUT4 in the presence of a maximally effective insulin concentration for the control group (pCIS2/GLUT4-HA). The actual value for the specific cell-associated radioactivity for the control group at 60 nM insulin was 592 ± 63 cpm. The best-fit curve generated from the mean data for the control group had an ED_{50} = 0.13 nM. The difference in the two curves is statistically significant by multivariate analysis of variance (F = 20.8, p = 0.00008).

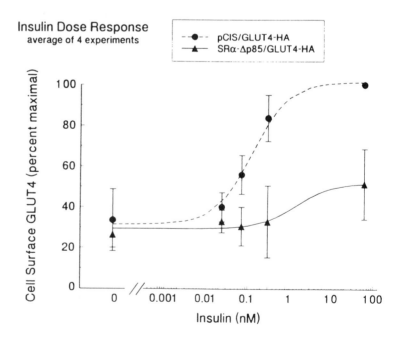

1994a]. Downstream signaling molecules such as phosphatidylinositol 3-kinase (PI3K) are activated when phosphotyrosine motifs on IRS-1 and other substrates bind to the SH2 domains of the downstream effectors. To determine if PI3K activity is necessary for insulin-stimulated translocation of GLUT4 in adipose cells we co-transfected cells with a dominant inhibitory mutant of the p85 regulatory subunit of PI3K (Δp85) along with GLUT4-HA; we then measured cell surface GLUT4-HA in response to insulin (Fig. 5.2) [Quon et al., 1995]. In control cells, insulin caused an approximately threefold increase in recruitment of epitope-tagged GLUT4 to the cell surface with an ED_{50} of approximately 0.13 nM. In the absence of insulin, cell surface GLUT4 was not significantly different between control cells and transfected cells expressing Δp85. However, cells transfected with Δp85 were markedly impaired in their ability to translocate epitope-tagged GLUT4 to the cell surface in response to insulin. That is, cell surface GLUT4 levels after insulin stimulation were not significantly different from basal levels. The striking ability of Δp85 to abolish insulin-stimulated translocation of GLUT4 in transfected adipose cells strongly suggests a necessary role for PI3K in insulin-stimulated glucose transport in insulin target tissues.

REFERENCES

Choi T, Huang M, Gorman C, Jaenisch R (1991): A generic intron increases gene expression in transgenic mice. Mol Cell Biol 11: 3070–3074.

Cushman SW, Salans LB (1978): Determinations of adipose cell size and number in suspensions of isolated rat and human adipose cells. J Lipid Res 19: 269–273.

Flier JS (1995): The adipocyte storage depot or node on the energy information superhighway? Cell 80: 15–18.

He Y, Chen H, Quon MJ, Reitman M (1995): The mouse obese gene. Genomic organization, promoter activity, and activation by CCAAT/enhancer-binding protein alpha. J Biol Chem 270: 28887–28891.

Hotamisligil GS, Spiegelman BM (1994): Tumor necrosis factor alpha: a key component of the obesity-diabetes link. Diabetes 43: 1271–1278.

Hotamisligil GS, Shargill NS, Spiegelman BM (1993): Adipose expression of tumor necrosis factor-alpha: direct role in obesity-linked insulin resistance. Science 259: 87–91.

Hotamisligil GS, Arner P, Caro JF, Atkinson RL, Spiegelman BM (1995): Increased adipose tissue expression of tumor necrosis factor-alpha in human obesity and insulin resistance. J Clin Invest 95: 2409–2415.

Ing BL, Chen H, Robinson KA, Buse MG, Quon MJ (1996): Characterization of a mutant GLUT4 lacking the N-glycosylation site: studies in transfected rat adipose cells. Biochem Biophys Res Commun 218: 76–82.

Karnieli E, Zarnowski MJ, Hissin PJ, Simpson IA, Salans LB, Cushman SW (1981): Insulin-stimulated translocation of glucose transport systems in the isolated rat adipose cell. Time course, reversal, insulin concentration dependency, and relationship to glucose transport activity. J Biol Chem 256: 4772–4777.

MacDougald OA, Lane MD (1995): Transcriptional regulation of gene expression during adipocyte differentiation. Annu Rev Biochem 64: 345–373.

Quon MJ, Zarnowski MJ, Guerre-Millo M, de la Luz Sierra M, Taylor SI, Cushman SW (1993): Transfection of DNA into isolated rat adipose cells by electroporation: evaluation of promoter #activity in transfected adipose cells which are highly responsive to insulin after one day in culture. Biochem Biophys Res Commun 194: 338–346.

Quon MJ, Butte AJ, Zarnowski MJ, Sesti G, Cushman SW, Taylor SI (1994a): Insulin receptor substrate 1 mediates the stimulatory effect of insulin on GLUT4 translocation in transfected rat adipose cells. J Biol Chem 269: 27920–27924.

Quon MJ, Guerre-Millo M, Zarnowski MJ, Butte AJ, Em M, Cushman SW, Taylor SI (1994b): Tyrosine kinase-deficient mutant human insulin receptors (Met1153 → Ile) overexpressed in transfected rat adipose cells fail to mediate translocation of epitope-tagged GLUT4. Proc Natl Acad Sci USA 91: 5587–5591.

Quon MJ, Chen H, Ing BL, Liu ML, Zarnowski MJ, Yonezawa K, Kasuga M, Cushman SW, Taylor SI (1995): Roles of 1-phosphatidylinositol 3-kinase and ras in regulating translocation of GLUT4 in transfected rat adipose cells. Mol Cell Biol 15: 5403–5411.

Reaven GM (1995): The fourth musketeer—from Alexandre Dumas to Claude Bernard. Diabetologia 38: 3–13.

Rodbell M (1964): Metabolism of isolated fat cells: effects of hormones on glucose metabolism and lipolysis. J Biol Chem 239: 375–380.

Spiegelman BM, Choy L, Hotamisligil GS, Graves RA, Tontonoz P (1993): Regulation of adipocyte gene expression in differentiation and syndromes of obesity/diabetes. J Biol Chem 268: 6823–6826.

Tontonoz P, Hu E, Spiegelman BM (1994): Stimulation of adipogenesis in fibroblasts by PPAR gamma 2, a lipid-activated transcription factor. Cell 79: 1147–1156.

Walston J, Silver K, Bogardus C, Knowler WC, Celi FS, Austin S, Manning B, Strosberg AD, Stern MP, Raben N, et al. (1995): Time of onset of non-insulin-dependent diabetes mellitus and genetic variation in the beta 3-adrenergic-receptor gene. N Engl J Med 333: 343–347.

Zhang Y, Proenca R, Maffei M, Barone M, Leopold L, Friedman JM (1994): Positional cloning of the mouse obese gene and its human homologue. Nature 372: 425–432.

APPENDIX: LIST OF MATERIALS AND SUPPLIERS

Materials	Catalogue Number	Su
Adenosine	102075	Boehringer-Biochem
Antibody, [I^{125}]-labeled sheep anti-mouse whole IgG (10 μCi/μg)	IM.131	Amersham
Antibody, 12CA5 anti-HA monoclonal	1666606	Boehringer-Biochem
Bacitracin	B0125	Sigma Chem
Bovine serum albumin, Bovuminar Cohn Fraction V, powder	3320; Lot P5503	Intergen Co
Collagenase type 1 359 U/mg	LS004196; Lot F5J576	Worthingto Corpora
Conical centrifuge tubes, 50 ml	25335-50	Corning Inc
Dinonylphthalate oil	210733	ICN Biome
DMEM, Dulbecco's modified Eagle's medium with 4.5 g glucose/l without glutamine	104	BioFluids, I

Dulbecco's phosphate buffered saline (PBS) with Ca^{2+} and Mg^{2+}	309	BioFluids, Inc.
Electroporation system:		
Gene Pulser II Apparatus	165-2105	BioRad Laboratories
Capatance Extender	165-2108	BioRad Laboratories
Pulse Controller	165-2100	BioRad Laboratories
Gene Pulser Cuvette 0.4 cm electrode gap (hand-fabricated)	165-2090	BioRad Laboratories
Filter unit 0.2 μm, 500 ml	156-4020	Nalge Co.
Filter unit 0.2 μm, 150 ml	155-0020	Nalge Co.
Filters, 250 μm mesh polymer	HC3-250	Apple Scientific Inc.
D-[U-^{14}C] Glucose, 294 mCi/mmol	NEC042X	DuPont NEN
HEPES sodium	H-7006	Sigma Chemical Company
Insulin		Sigma
Magnetic stirring bars $\frac{5}{8} \times \frac{5}{16}$ inch (L $\times$ D)	F37120-7836	Apple Scientific Inc.
(R)-N^6-(1-methyl-2-phenylethyl)adenosine (PIA)	P4532	Sigma Chemical Company
Microcentrifuge tubes, polyethylene, 4 mm $\times$ 45 mm	F19928	Apple Scientific Inc.
Petri dish, 60 $\times$ 15 mm	3002	Becton Dickinson Labware
PIA ((R)-N^6-(1-methyl-2-phenylethyl)adenosine)	P4532	Sigma Chemical Company
Pipette tips, 200 μl, wide bore	410006	Stratagene
Polyethylene microcentrifuge tubes, 4 mm $\times$ 45 mm	F19928	Apple Scientific Inc.
Polyethylene scintillation vials, 7 ml	125500	Research Products Inc.
Polyethylene scintillation, vials, 20 ml	986704	Apple Scientific Inc.
Polypropylene centrifuge tube, 50 ml	25335-50	Corning Incorporated
Polypropylene round bottom tube, 17 $\times$ 100 mm	2059	Becton Dickinson Labware
Polypropylene vial, 30 ml, low density	2003-0001	Nalge Co.
Polystyrene centrifuge tubes, 15 ml	25311-15	Corning Incorporated
Rats Sprague-Dawley (CD strain), male		Charles River Breeding Laboratories
Wizard Megaprep kit	A7300	Promega
Wintrobe sedimentation Cannula	063-578	Apple Scientific Inc.

6

Calcium Phosphate Transfection

K. J. Conn, A. Degterev, M. R. Fontanilla, C. B. Rich, and
J. A. Foster

Department of Biochemistry, Boston University School of Medicine, 80 E.
Concord St., Boston, Massachusetts 02118

DNA Transfer to Cultured Cells, Edited by Katya Ravid and R. Ian Freshney.
ISBN 0-471-16572-7   © 1998 Wiley-Liss, Inc.

1. INTRODUCTION TO CALCIUM PHOSPHATE TRANSFECTION

The ability to transfer exogenous DNA into cultured mammalian cells has revolutionized the way in which genes are studied. Arguably one of the most dramatic illustrations of the power of this technology was the transformation of NIH 3T3 cells after the introduction of DNA isolated from chemically transformed cells [Shih et al., 1979]. This experiment, demonstrating the ability of cellular oncogenes to pass a transformed phenotype to normal cells via genomic DNA, was accomplished using the calcium phosphate co-precipitation method developed by Graham and Van der Eb [1973].

Seeking to improve upon current methods available to them, Graham and Van der Eb took advantage of the observation that divalent cations such as calcium and magnesium promoted the uptake of DNA into bacterial cells [Spizizen et al., 1966]. They found when DNA was precipitated with calcium phosphate prior to introduction in culture, the efficiency of transfection was improved by as much as 100-fold over the DEAE-dextran method. In the more than 20 years since its conception, the calcium phosphate method of DNA transfection has proven to be a useful method for both transient and stable expression of exogenous DNA. The protocol described in this chapter is an adaptation of the original protocol described by Graham and Van der Eb [1973]. It has been optimized for transiently expressed DNA and includes modifications introduced by Chen and Okayama [1987], as well as Ritzenthaler et al. [1991].

1.1. Advantages and Disadvantages

Calcium phosphate transfection offers many advantages over other methods of DNA transfer. In addition to being a relatively inexpensive procedure, the calcium phosphate method can accommodate high concentrations of DNA. Using this method one can introduce as much as 60 μg of DNA per 100-mm dish, whereas the DEAE-dextran technique loses efficiency at DNA concentrations greater that 1 μg/60-mm dish [Howard et al., 1971; Warden and Thorne, 1968]. Therefore, when assessing the expression of transfected DNA that its cellular counterpart is intrinsically expressed at low levels, the calcium phosphate method may be preferred over the DEAE-dextran method [Graham and Van der Eb, 1973]. Another strength of calcium phosphate transfection is that cells tend to pick up a representative sampling of DNA in the precipitate. Hence one can prepare a 10:1 ratio of two unique plasmids and expect those plasmids to be present in that ratio in the transfected cells.

The calcium phosphate method may not be ideal for all experimental conditions. The efficiency of transfection is often no greater than 10% [Somasundaram et al., 1992] and can vary dramatically depending on the cell type and experimental conditions. An additional drawback to the calcium phosphate method is that calcium phosphate disrupts the cell membrane and can lead to cell death [Wigler et al., 1977, 1978]. Therefore, the calcium phosphate method may not be optimal for experimental protocols that require an extended incubation period of DNA with cells.

1.2. Mechanism of Action

How the precipitation of DNA with calcium phosphate facilitates the expression of exogenous DNA is unclear; however, the mechanism of action probably consists of two stages: adsorption, followed by uptake [Graham and Van der Eb, 1973]. Following the addition of calcium phosphate DNA coprecipitate to cells in culture, the coprecipitate sediments and becomes adsorbed to the cell membrane. Adsorption facilitates uptake by increasing the effective DNA concentration near the cell membrane.

After adsorption of co-precipitate, DNA is taken up by the cell via a calcium-requiring process [Graham and Van der Eb, 1973]. The details of uptake are not known; however, it is likely that endocytosis is a major route for entry, as the treatment of the cells with chloroquine (which has been shown to prevent degradation of proteins absorbed by endocytosis) significantly improves the efficiency of transfection [Luthman and Magnusson, 1983]. It is unclear whether DNA is released from the calcium phosphate precipitate prior to uptake, or alternatively, if particles of calcium phosphate-DNA complex are taken up together and the DNA is subsequently released [Graham and Van der Eb, 1973]. Ultimately, once inside the cell, some of the DNA evades destruction and makes its way to the nucleus, where it can be transiently expressed and/or integrated into the mammalian chromosome.

2. FACTORS INFLUENCING CALCIUM PHOSPHATE-DNA PRECIPITATE FORMATION

Although calcium phosphate-DNA co-precipitates form spontaneously in supersaturated solutions, highly effective precipitates for transfection purposes can be generated only in a very narrow range of physicochemical conditions controlling the initiation and growth of the precipitate complex [Jordan et al., 1996]. In general, the highest transfection efficiencies correlate with the generation of many very small, fine particles, while avoiding big crystals of calcium phosphate-DNA co-precipitate (see Fig. 6.1). There are several parameters that influence precipitate formation including DNA concentration, temperature, time of co-precipitate incubation prior to addition to cells, and pH of the HEPES-buffered saline (HBS) buffer.

The concentration of DNA used can influence the quality of precipitate formed. Chen and Okayama [1987] observed that by increasing the concentration of DNA from 10 to 30 μg/100-mm dish a transition from a coarse to a fine precipitate occurred and that this correlated with an increase in transfection efficiency. Increasing the amount of soluble DNA in a supersaturated solution of calcium and phosphate prolongs the period during which a precipitate forms and allows for the slow formation of a fine precipitate. If too high in concentration, however, DNA may prevent precipitate formation entirely [Jordan et al., 1996].

Temperature also affects precipitate formation. The solubility of calcium phosphate is higher at lower temperatures. Most protocols

Fig. 6.1. Visualization of fine and course calcium phosphate-DNA co-precipitate under a light microscope. In general, highest transfection efficiency correlates with the generation of many very small, fine particles of calcium phosphate-DNA co-precipitate **(A)**, while avoiding big crystals **(B)**. The photographs are with 100× magnification.

Conn et al.

recommend that precipitate formation be carried out at room temperature; however, strict temperature control may optimize results.

How long the co-precipitate is incubated prior to introduction into culture can influence the quality of the precipitate as well. In theory, almost all of the soluble DNA in the reaction mix can be bound into an insoluble complex in less than 1 min. Extending the reaction time to 20 min or more often results in aggregation and/or growth of particles, which can reduce the level of gene expression. Most protocols recommend letting the DNA and calcium phosphate react for no more than 20 min prior to the addition to cells; however, this may vary depending on the type of cell being transfected and experimental conditions. For example, Graham and Van der Eb [1973] found that incubation for 20 to 60 min was optimal.

Finally, the pH of the HBS buffer can greatly influence precipitate formation. Graham and Van der Eb [1973] found the optimum pH range for transfection was between 7.05 and 7.12. Below pH 6.9 no precipitate was observed, at pH 7.0 a very fine precipitate formed, and with increasing pH the precipitate formed more rapidly and was heavier and coarser. In contrast, for the protocol described by Chen and Okayama, [1987] the optimum pH for transformation was around 6.95, and at pH 7.05–7.15 the transformation frequency was several times lower.

3. FACTORS INFLUENCING TRANSFECTION EFFICIENCY

Of the factors influencing transfection efficiency the formation of quality precipitate is of paramount importance; however, there are several other factors which can also influence the efficiency of transfection. These include the amount of DNA used for transfection, the length of time the precipitate is left on the cell, and the use and duration of a glycerol or DMSO shock.

In addition to influencing the integrity of calcium phosphate-DNA co-precipitate, the total amount of DNA introduced to a cell can influence the efficiency of uptake. In some cell lines introducing more than 10–15 μg of DNA into a 100-mm dish can result in excessive cell death. In contrast, many primary cells need a high concentration of DNA to see biological activity of the exogenous DNA [Selden, 1987].

How long the precipitate is in culture can also influence the efficiency of transfection. An incubation period of as little as 4 h can allow the precipitate to associate with the plasma membrane; however, most protocols recommend culturing the cells with the precipitate at least overnight. All cells may not tolerate an overnight incubation, as calcium phosphate can be toxic. Consequently, most protocols advise not leaving the precipitate on the cells for more than 24 h.

Finally, many protocols recommend the use of a glycerol or DMSO shock at the end of the period during which the precipitate and cells are incubated. The brief incubation of cells with glycerol often increases transfection efficiency, possibly by affecting the permeability of the membrane. Glycerol shocking also helps to remove any remaining calcium phosphate-DNA coprecipitate that was not taken up by the cell, and does so more effectively than washing alone. Although it is helpful with some cell types, the shock may not always improve transfection efficiency and may lead to the removal of cells from the tissue culture surface and/ or cell death. Using DMSO in place of glycerol is somewhat less harmful to the cells, but does not seem to work as well [Kingston, 1987].

4. PREPARATION OF DNA FOR TRANSFER

Successful expression of foreign DNA in targeted cells ultimately depends on the integrity and purity of the DNA used for transfer. Impurities in the DNA preparation can be deleterious to transfection efficiency. For example, many proteins from *Escherichia coli* that are employed to generate plasmid DNA preparations are toxic to mammalian cells; therefore, plasmid DNA should be purified by two successive CsCl gradient centrifugations to remove contaminating proteins, followed by two rounds of DNA precipitation to get rid of contaminating CsCl, which also can decrease transfection efficiency. Additionally, care should be taken to ensure that all ethidium bromide used during plasmid preparation is completely removed. Direct exposure to solutions containing ethidium bromide, as well as vigorous handling of DNA, can cause nicking and relaxation of DNA. Relaxed DNA is much less transfectable than supercoiled DNA. It is recommended that prior to transfection, DNA should be visualized after separation on an agarose gel to insure that most of it is in the supercoiled form.

All DNA used for transfection should be resuspended in sterile water. Be careful to avoid resuspending the DNA in any buffer containing Tris, as Tris may alter the pH of the precipitate and therefore reduce transfection efficiency. When culturing DNA with cells for an extended period of time, the recommended procedure is to precipitate the DNA first with ethanol, then to air dry the pellet in a tissue culture hood before resuspending it in sterile water.

5. CULTURE OF CELLS FOR TRANSFECTION

In addition to the previously mentioned factors, the density at which the cells are seeded as well as the CO_2 levels in which cells are cultured can influence the efficiency of transfection. In general, cells cultured as a subconfluent monolayers (60–90% confluent) give higher efficiencies of transfection than very dense cultures [Graham and Van der Eb, 1973]. Multiple studies have shown that transfection efficiency is significantly higher using actively growing cells rather than quiescent cells [Somasundaram et al., 1992; Strain et al., 1985; Kjer and Fallon, 1991]. Together these observations suggest a requirement for cell division after addition of DNA for highest expression. It is important however, that cells are not seeded too sparsely, as a wide range of phenotypic features depends on the ability of cells to form cell–cell contacts. As a rule, the optimal density is one that produces a near confluent dish when the cells are harvested.

Lastly the CO_2 level in which cells are cultured can influence transfection efficiency. It may be that the optimal CO_2 concentration for transfection differs from that used for routine passage and culture. For example, Chen and Okayama [1987] found that lowering the CO_2 from 5% to 3% during the incubation of the calcium phosphate-DNA coprecipitate improved transfection efficiency. Lowering of the CO_2 level causes the pH of the culture media to increase, although how this leads to increased transfection efficiency is unclear. It is recommended that during the transfection period the media should be between pH 7.2 and 7.4 and that the CO_2 concentration be kept close to 5% [Kingston, 1987].

6. PREPARATION OF REAGENTS

The following reagents are required for the preparation of DNA for transfer. Reagents are prepared using sterile H_2O and/or are sterilized by filtration through 0.2-μm filters. Once filtered, aseptic technique should be observed, where appropriate, to prevent contamination, both in the preparation and subsequent handling.

Sterile Reagents

HEPES-buffered saline (HBS), 2×
NaCl .... 16.3 g/l (280 mM)
HEPES .... 11.9 g/l (50 mM)
Na_2HPO_4 .... 0.4 g/l (2.82 mM)

Dissolve the reagents in autoclaved H_2O
Adjust the pH to 7.12
Filter-sterilize through 0.2-μm filter
Store at −20°C
May be stored for up to 1 month
Note: An exact pH is extremely important for efficient transfection. The pH of the solution can change during storage. In our hands freshly made 2× HBS works better than defrosted solutions for precipitate formation, however defrosted HBS may be used for preparation of the 15% Glycerol solution.

2 M CaCl₂
$CaCl_2$ (dihydrate), 2 M
Dissolve in autoclaved H_2O
Sterilize by filtration through a 0.2-μm filter and store −20°C.
May be stored for up to 1 month.
Note: Some investigators feel that certain lots of $CaCl_2$ yield much higher transfection efficiencies than others. In our hands freshly prepared $CaCl_2$ works better than defrosted solutions.

Puck's saline
NaCl .... 8.0 g/l   (137 mM)
KCl .... 0.4 g/l   (5.37 mM)
KH_2PO_4 .... 0.15 g/l   (1.10 mM)
$Na_2HPO_4 \cdot 7 H_2O$ .... 0.29 g/l   (1.08 mM)
Glucose .... 1.1 g/l   (6.10 mM)

Adjust to pH 7.4
Filter-sterilize through 0.2-μm filter
Store at 4°C
May be stored for up to 6 months
Note: PBSA can be used in place of Puck's saline.

15% Glycerol in HBS
Glycerol .... 15% v/v
H_2O .... 35% v/v
HBS, 2× .... 50% v/v

Filter-sterilize through a 0.2-μm filter
Prepare fresh each time

0.25 M Tris HCl
Tris .... 0.25 M

Adjust to pH 8.0 with HCl
Store at room temperature
May be stored for up to 6 months

TEN, 10×
Tris, pH 8.0 .... 100 mM
EDTA, pH 8.0 ...... 10 mM
NaCl .... 1 M

Store at room temperature
May be stored for up to 6 months

Sterile Materials

Tubes, 15 ml
Conical tubes, 50 ml
Pasteur pipettes, 9″
Eppendorf tubes
Cell scrapers
Tissue culture petri dishes, 100 mm
Note: The use of plastic pipettes is preferred during DNA-calcium phosphate co-precipitate formation, as DNA sticks to glass and any particulate matter introduced on glassware may seed large crystal growth during the precipitation step.

7. TRANSFECTION PROTOCOL

The step-by-step protocol described below has been optimized for the transient analysis of chimeric promoter chloramphenicol acetyltransferase (CAT) and β-galactosidase (β-gal) constructs in primary rat aortic smooth muscle cells. An example of the time course of a transfection experiment is given below. This will vary depending on the specific experimental design.

Day 1	Day 2	Day 3	Day 4	Day 5	Day 6
AM	AM	AM	AM	AM	
Pass cells	Feed cells				Assay for reporter activity
		PM	PM	PM	
	PM	Glycerol	Add growth	Harvest cells	
	Add DNA	shock	factor		

To ensure that sample batches transfected at different times have transfection efficiencies as similar as possible, it is important to perform all steps as uniformly as possible: for example, time the rate of mixing of DNA-calcium phosphate and HBS solutions (time between drops), add precipitate to cells after a predetermined time of incubation, maintain a constant flow of air for all samples, and add precipitate to cells at a similar rate. Additionally, all cell handling steps should be done in a tissue culture hood to avoid contamination, and all solutions applied to cells should be prewarmed to 37°C or to at least room temperature.

Protocol 7

Day 1

(a)   Cells are passed into 100-mm dishes in 10% serum

Note: Cell densities and the percentage of serum will depend on the cell type and experimental conditions.

Day 2

(a)   Cells should be fed with complete medium 2–4 h prior to addition of precipitate.

(b)   Add the following to a 15-ml tube:

   Experimental plasmid .... 15–30 μg/dish
   β-galactosidase plasmid ....   5 μg/dish
   Sterile H$_2$O to .... 500 μl/dish
   CaCl$_2$, 2 M .... 62 μl/dish

(c)   Add the above solution no faster than 1 drop per second to a 50-ml tube containing a volume of 2× HBS equal to: 500 μl/ 100-mm dish with air to prevent precipitate from clumping (see Fig. 6.2).

(d)   Allow the precipitate to stand at room temperature for 10–20 min.

(e)   Resuspend the precipitate by pipetting and add 1 ml of precipitate solution onto each 100-mm dish of cells, swirling evenly to cover the cell layer.

(f)   Incubate overnight.

Note: After mixing of the DNA-CaCl$_2$ solution with the HBS solution a slight opacity appears within a few minutes, indicating that a precipitate has been formed.

   Depending on how efficiently the cell type you are working with transfects, you may wish to use more than one plate per experimental condition. For example in our laboratory we prepare in the same tube enough DNA and CaCl$_2$ to transfect three dishes, then combine the cells from these dishes at time of harvest. Importantly, do not attempt to precipitate more than three plates worth of DNA at a time as the quality of the precipitate suffers.

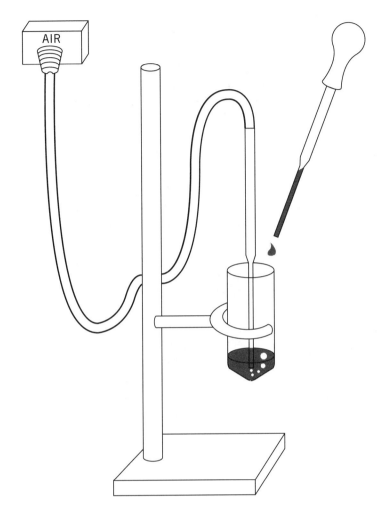

Fig. 6.2. Formation of the calcium phosphate-DNA coprecipitate. The DNA/CaCl$_2$ solution is added dropwise to the 2× HBS in which air is forcefully percolating. A filter (not shown) may be put between the air source and the pipette to ensure sterility.

We perform this step outside of a tissue culture hood, as we regularly use antibiotics in the media of our cultured cells and culture the DNA for a short period of time. This step may be performed in a tissue culture hood to ensure sterility as needed.

Variations of this step include reversing the order of addition of the two solutions. For example, instead of adding the DNA-CaCl$_2$ to the HBS as described above, some alternative protocols add the HBS to the DNA-CaCl$_2$ solution.

CaCl$_2$ should be added to DNA just prior to mixing with HBS and not ahead of time.

Air flow should be as strong as possible, while avoiding splattering of reagents outside the 50-ml tube.

Day 3
(a) After at least 16 h of transfection, remove media from cells.
(b) Wash the cell layer two times with PBSA or Puck's saline.

Calcium Phosphate Transfection   121

(c) Add 3 ml of the 15% glycerol in HBS per dish, swirling to cover the cell layer.

(d) Leave the glycerol solution on for 30 s, then remove.

(e) Wash the cell layer two times with PBSA or Puck's saline.

(f) Add to the cells the appropriate media for experimental conditions.

(g) Incubate for desired time.

Note: When shocking, it is important that glycerol not be in contact with the cells more than 30 s, as excessive exposure to the glycerol solution can result in cell death. To minimize the time of exposure, the first wash of Puck's saline after the shock can be added before removal of the glycerol. This dilutes the glycerol, thus minimizing exposure time.

After shocking, cells are only tenuously attached to the plate; therefore it is important to be gentle with the cells. For example, slowly dribble the media down the sides of the dish to avoid dislodging cells from the plate.

The media will become cloudy with precipitate during the overnight incubation period; however, appearance of excessive cloudiness accompanied by abnormal odor, cell death, and moving particles may indicate that the cells have become contaminated and should not be used.

For transient analysis, cells should be harvested 24–48 h (but no longer than 72 h) after shock, as transiently transfected vectors begin to degrade.

Day 4

(a) After 24 h, either change the media according to experimental protocol or harvest cells as described for day 5.

Day 5 (Harvesting Cells)

(a) Remove the medium by aspiration.

(b) Wash the cell layer twice with Puck's saline or PBSA.

(c) Scrape the cells in 500 μl 1× TEN (dilute 10× TEN with sterile H_2O) and put them into a labelled Eppendorf tube on ice.

(d) Repeat (c) with an additional 500 μl 1× TEN and combine with the cells in the Eppendorf tube.

(e) Pellet the cells in a microcentrifuge for 3 min at high speed (16,749 g).

(f) Discard the supernatant.

(g) Resuspend the cell pellet in 150 μl 0.25 M Tris HCl, pH 8.0.

(h) Freeze the cells at −20°C.

(i) Defrost cells on ice prior to sonication.

(j) Centrifuge the cells for 10 min at high speed (16,749 g) at 4°C.

(k) Transfer the supernatant to new tube and discard the pellet.

(l) Store at −20°C

Note: The size of the cell pellet can vary depending on the number of cells used in the assay. If the pellet is larger than ~50 μl, the amount of Tris should be increased accordingly.

Cells should be sonicated for no more than 10 s at a time and no more than four times in a row before being placed on ice to avoid overheating. Sonication should be performed until the cell extract becomes homogeneous with no apparent clumps.

Repeated freezing and thawing of samples should be avoided, as freezing may lead to proteins falling out of solution and can decrease reporter activity.

8. ASSAY OF REPORTER PRODUCTS

Depending on the experimental design, reporter constructs may be used to assess the biological activity of exogenous DNA. There are several reporters that can be used with the calcium phosphate transfection method, including chloramphenicol acetyltransferase (CAT) (see Chapter 10), β-galactosidase (β-gal) (see Chapters 7 and 9), luciferase (LUC), and human growth hormone (hGH). Our laboratory uses CAT as the reporter of interest and β-gal for normalization. We get the most consistent results when equal volumes of cell lysate are assayed for β-gal activity on the day of harvest, followed by the assay of equal amounts of protein for CAT activity on the same day that protein concentration is determined.

ACKNOWLEDGMENTS

This work was supported by National Institute of Health (NIH) Grants HL-46902 and HL13262 and Training grant AG00115 (K. J. Conn).

REFERENCES

Chen C, Okayama H (1987): High-efficiency transformation of mammalian cells by plasmid DNA. Mol Cell Biol 7: 2745–2752.

Graham FL, Van der Eb AJ (1973): A new technique for the assay of infectivity of human adenovirus 5 DNA. Virology 52: 456–467.

Howard BV, Estes MK, Pagano JS (1971): The uptake of SV-40 DNA by nonpermissive cells in the presence of DEAE-dextran. Biochem Biophys Acta 228: 105–116.

Jordan M, Schallhorn A, Wurm FM (1996): Transfecting mammalian cells: optimization of critical parameters affecting calcium-phosphate precipitate formation. Nucleic Acids Res 24: 596–601.

Kingston RE (1987): Transfection of DNA into eukaryotic cells. In Ausubel, FM et al. (eds): "Current Protocols in Molecular Biology." New York: Wiley Interscience, pp 9.1.1–9.1.4.

Kjer KM, Fallon AM (1991): Efficient transfection of mosquito cells is influenced by the temperature at which DNA-calcium phosphate co-precipitates are prepared. Arch Insect Biochem Physiol 16: 189–200.

Luthman H, Magnusson G (1983): High efficiency polyoma transfection of chloroquine treated cells. Nucleic Acids Res 11: 1295–1309.

Ritzenthaler JD, Goldstein RH, Fine A, Liehtler A, Rowe DW, and Smith BD (1991): Transforming-growth-factor-β activation elements in the distal promoter regions of the rat α1 type I collagen gene. Biochem J 280: 157–162.

Selden RF (1987): Optimization of transfection. In Ausubel FM et al. (eds): "Current Protocols in Molecular Biology." New York: Wiley Interscience, p 9.4.2.

Shih C, Shilo B-Z, Goldfarb MP, Dannenberg A, Weinberg RA (1979): Passage of phenotypes of chemically transformed cells via transfection of DNA and chromatin. Proc Natl Acad Sci USA 76: 5714–5718.

Somasundaram C, Tournier I, Feldmann G, Bernuau D (1992): Increased efficiency of gene transfection in primary cultures of adult rat hepatocytes stimulated to proliferate: a comparative study using the lipofection and the calcium phosphate precipitate methods. Cell Biol Internatl Reports 16: 653–662.

Spizizen J, Reilly BE, Evan AH (1966): Microbial transformation and transfection. Annu Rev Microbiol 20: 371–400.

Strain AJ, Wallace WA, Wylie AH (1985): Enhancement of DNA-mediated gene transfer by high Mr carrier DNA in synchronized CV-1 cells. Biochem J 225: 529–533.

Warden D, Thorne H V (1968): The infectivity of polyoma virus DNA for mouse embryo cells in the presence of diethlaminoethly-dextran. J Gen Virol 3: 371–377.

Wigler M, Silverstein S, Lee LS, Pellicer A, Cheng YC, Axel R (1977): Transfer of purified herpes virus thymidine kinase gene to cultured mouse cells. Cell 11: 223–232.

Wigler M, Pellicer A, Silverstein S, Axel R (1978): Biochemical transfer of single-copy eucaryotic genes using total cellular DNA as donor. Cell 14: 725–731.

APPENDIX: LIST OF MATERIALS AND SUPPLIERS

Item	Supplier	Catalogue Number
$CaCl_2$ (dihydrate, cell culture grade)	Sigma	C-7902
EDTA (electrophoresis grade)	Fisher	BP118-500
Glucose	Sigma	G-8270
Glycerol (molecular biology grade)	IBI Inc	IBI5750
HEPES (molecular biology grade)	Sigma	H-0891
KCl (cell culture grade)	Sigma	P-5405
KH_2PO_4	Fisher	P-382
Na_2HPO_4 (anhydrous, cell culture grade)	Sigma	S-5136
$Na_2HPO_4 \cdot 7\ H_2O$	Sigma	S-9390
NaCl (cell culture grade)	Sigma	S-5886
Tris (molecular biology grade)	Fisher	BP152-5

7

Optimization of Calcium Phosphate Transfection for Reporter Gene Analysis

John V. O'Mahoney and Timothy E. Adams

The Children's Medical Research Institute, Locked Bag 23, Wentworthville, NSW 2145, Australia (J.V.O'M). CSIRO, Division of Molecular Science, 343 Royal Parade, Parkville, Victoria 3052, Australia (T.E.A.).

DNA Transfer to Cultured Cells, Edited by Katya Ravid and R. Ian Freshney.
ISBN 0-471-16572-7 © 1998 Wiley-Liss, Inc.

1. INTRODUCTION

1.1. Reporter Gene Analysis

Reporter genes encoding β-galactosidase (β-gal), secreted alkaline phosphatase, human growth hormone, chloramphenicol acetyltransferase (CAT), luciferase, β-glucuronidase (GUS), and green fluorescent protein (GFP) [reviewed by Alam and Cook, 1990; Ausubel et al., 1995] have played a pivotal role in deciphering the complexities of mammalian gene expression. The capability of specific *cis*-acting DNA sequences to regulate gene expression can be assessed by subcloning putative regulatory elements into reporter gene expression vectors that are deficient in promoter and/or enhancer elements. Putative promoter elements are subcloned into promoterless reporter plasmids upstream of the reporter gene, while enhancer or repressor elements are positioned 5' or 3' to the reporter gene in a plasmid bearing a functional minimal promoter. These are then introduced into mammalian host cells for expression of the reporter gene. As the protein encoded by the reporter gene is not normally expressed in the host cells, and has well-established simple, and sensitive detection methods, it provides a convenient means by which to measure the level of expression imparted by the sequences in question. Although the reporter protein assay is an indirect measure of transcriptional activation, it is a practical way by which to avoid the laborious practice of quantitating RNA levels directly (e.g., by Northern analysis or RNase protection assay).

The introduction of reporter gene constructs into host cells may be achieved by transgenesis, stable transfection, or transient transfection. The transgenic approach is technically demanding, expensive, and time-consuming, but once established, it offers the advantage of a true *in vivo* system, in which the reporter construct is chromosomally integrated and subject to the full repertoire of transcriptional machinery, much of which may be lost in a cell culture system. The lack of a suitable cell culture system sometimes makes the transgenic approach unavoidable. For example, investigations into the regulatory elements confining muscle-specific gene expression to particular muscle fiber subtypes relies on transgenic models, as functional innervation is considered to play an important role in muscle development [Levitt et al., 1995]. The fact that the genotype is heritable with both transgenic and stable transfection approaches eliminates the variables associated with introducing the reporter construct into the cells each time an experiment is performed. Both approaches, however, are compromised by the possibility that the site of integration adversely affects the function of the regulatory sequence or of the cell.

1.2. Limitations on the Use of Internal Control Plasmids for Transient Transfections

If a suitable cell culture system is available, the transient transfection of reporter genes is commonly used for the initial characterization of transcriptional regulatory elements. The influence of a putative *trans*-acting factor may be assessed by transiently cotransfecting a reporter gene construct containing the *cis*-acting element with an expression construct for the factor. The influence of hormonal and other cell culture conditions may also be evaluated in cells transiently transfected with a reporter gene. The reliability of the data generated from such experiments relies on the strength of the expression construct(s), and the ability to make valid comparisons between different sets of transfections. Both criteria depend on the ability to achieve consistent and efficient transfections, or at least to correct for any variability in transfection efficiency. It is therefore common practice to cotransfect an internal control plasmid that encodes for a second reporter protein under the control of a heterologous, ubiquitously active promoter. The activities of replicate transfections of the test plasmid (often with two or more independent plasmid preparations) are then corrected for transfection variability based on the relative activities of the cotransfected internal control, and the results are averaged.

It is usually assumed that the use of a co-transfected internal control plasmid adds credibility to the data generated from transient transfection experiments. This could, however, be an oversimplification. In a study of the upstream regulatory regions of the human papillomavirus, Farr and Roman [1992] discovered that the activity of their internal control plasmid, pRSV-β-gal (a β-galactosidase reporter plasmid under the control of the Rous sarcoma virus promoter/enhancer regions) varied according to which test plasmid was co-transfected. Similarly, Bergeron and colleagues noted that the co-transfection of specific test plasmids with β-gal internal control plasmids resulted in significant differences in β-gal expression levels [Bergeron et al., 1995]. Therefore, in both studies, using the β-gal internal control activities to correct their test plasmids' activities for variations in transfection efficiencies would generate false and misleading data. We also have evidence that the co-transfection of internal control plasmids may influence the activity of test reporter constructs. In an investigation of the upstream regulatory region of a liver-specific promoter for the growth hormone receptor, we have sometimes been unable to detect reporter activity from the test plasmid. This occurred even though there was evidence that the cells were being transfected with reasonable efficiency based on the activities of an internal control plasmid, pSV-β-gal, and a positive control plasmid, pGL2-Control (β-galactosidase and luciferase reporter plasmids, respectively, which are under the control of SV40 promoter/ enhancer regions). It was subsequently discovered that if we omitted the β-gal internal control plasmid from the transfections, we were able to detect luciferase activity from the test plasmid. Similar results were observed with another β-gal internal control reporter plasmid that was under the control of the cytomegalovirus promoter/enhancer regions (pCMV-β-gal). In Fig. 7.1 it can be seen that both pSV-β-gal and pCMV-β-gal, when used as internal control plasmids, adversely influence luciferase reporter activity conferred by the growth hormone receptor regulatory region in a dose-dependent manner.

The results of these studies make it apparent that, in transient transfections, co-transfected internal control plasmids should be used with caution to correct for variations in transfection efficiency. We do not know the precise nature of the interaction between different co-transfected reporter genes that causes variation in their activities. However, it is likely to be a transcriptional effect mediated by the competition of common *cis*-acting elements within the cotransfected plasmids for limiting transcription factors, whether these be repressors or activators. Alternatively, it may be an indirect effect whereby the titration of one or more factors subsequently influences the abundance of other *trans*-acting factors. A common approach in delineating the regulatory regions of a gene is to compare the level of reporter activity conferred by various deletions or mutations. In a situation where cotransfected reporter activities are influenced by the titration of *trans*-acting factors, the deletion or mutation of sequences within the test plasmid could have a significant impact on the activity conferred by the internal control plasmid.

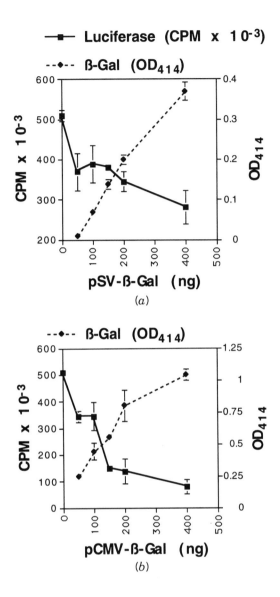

Luciferase (CPM x 10⁻³)

ß-Gal (OD₄₁₄)

(a) pSV-ß-Gal (ng)

ß-Gal (OD₄₁₄)

(b) pCMV-ß-Gal (ng)

Fig. 7.1. Inhibitory effect of β-gal internal controls on a luciferase reporter gene's activity. HuH7 cells cultured in 4-well plates were transfected with 1.6 µg of pGHR-GL2 (a luciferase reporter plasmid under the control of a growth hormone receptor promoter) and increasing amounts (0–400 ng) of either pSV-β-gal (**A**) or pCMV-β-gal (**B**). The total amount of plasmid transfected into each well was adjusted to 2.0 µg with pBluescript (Stratagene), a basic cloning vector. Each well was assayed for β-gal activity (OD₄₁₄) and luciferase activity (cpm × 10⁻³). The mean ± SD for four independent transfections are shown. Reproduced from O'Mahoney and Adams [1994], with permission from the publisher.

It is incorrect to assume that the activity of the internal control is unaffected when analyzing the transcriptional effects of various hormones on reporter constructs. Similarly, it is conceivable that the expression of the internal control will be affected when examining the influence of exogenous transcription factors by the co-transfection of transcription factor expression plasmids. We have observed both such effects in our studies with growth hormone receptor promoter reporter constructs, pCMV-β-gal internal control plasmid, dexamethasone, and members of the C/EBP family of transcription factors (unpublished observations). These effects could be due to direct interactions with the internal control's regulatory elements, secondary influences due to an alteration in the expression of the cell's repertoire of transcription factors by the hormones/transcription factors, or indirect effects due to protein–protein interactions. Therefore, internal control plasmids should be relied upon only for cells cultured under identical conditions, and only when it can be demonstrated that they do not affect the activities of the test-reporter genes. Although the procedure is not ideal, if internal control plasmids do influence the activity of the test plasmids, it may be feasible to perform dose–response experiments similar to those in Fig. 7.1, and to transfect limiting amounts of the internal control so as to minimize their impact on the test plasmid's activities.

An alternative approach to correcting the activity of the test plasmid for transfection efficiency is to perform a Hirt assay [Hirt, 1967; Ravid et al., 1991]. This avoids the use of a cotransfected internal control plasmid and relies on the ability to recover the transfected test plasmid's DNA from the cell lysates. The recovered plasmid is then quantitated by Southern blot analysis and used to normalize the test plasmid's reporter activity.

As the sole purpose of internal control plasmids is to correct for variations in the efficiency of transient transfections, optimizing the transfection procedures so as to minimize any variability would make them unnecessary. Some of the critical factors that influence the efficiency and ability to transiently transfect cells by the calcium phosphate coprecipitation method are described below.

2. PREPARATION OF MEDIA AND REAGENTS

Prepare all solutions with analytical or tissue culture grade chemicals in pure deionized (e.g., Millipore's Milli-Q) or double-distilled water (UPW, see Preface). Wherever possible, use disposable plasticware, or clean glassware that is dedicated to tissue culture.

Terrific broth
Tryptone .... 12 g
Yeast extract .... 24 g
Glycerol .... 4 ml
UPW .... to 900 ml

Sterilize by autoclaving.
When cool, add 100 ml of sterile (autoclaved) salt solution:

KH_2PO_4 ...... 170 mM
K_2HPO_4 ...... 720 mM

2× BES-buffered saline (2× BBS)

N,N-bis(2-hydroxyethyl)-2-aminoethanesulfonic

acid (BES) .... 50 mM

NaCl .... 280 mM

Na_2HPO_4 .... 1.5 mM

Adjust to pH 6.96 or 7.02 with 1 N NaOH.
Filter sterilize through a 0.22-μm nitrocellulose filter.
Store aliquots at $-20°C$.

As described in the transfection protocol, the pH is critical. Given the accuracy limitations of laboratory pH meters and day-to-day variability, it is recommended that preparations at both pH 6.96 and 7.02 be kept on hand. The best solution on any given day can then be chosen with reference to the standing time (see Section 6). The pH of these solutions can drift over time (>2 months).

$CaCl_2$, 2.5 M

Filter-sterilize 2.5 M $CaCl_2$ through a 0.22-μm nitrocellulose filter and store aliquots at $-20°C$.

Glutaraldehyde solution

Na_2HPO_4 .... 60 mM

NaH_2PO_4 .... 40 mM

$MgCl_2$ .... 1 mM

Glutaraldehyde .... 1% v/v

X-gal (4-Cl-5-Br-3-indolyl-β-galactoside) solution

The following solution (without X-gal) is prepared and stored at 4°C:

Sodium phosphate, pH 7.0 .... 10 mM

$MgCl_2$ .... 1 mM

NaCl .... 150 mM

$K_4Fe(CN)_6 \cdot 3H_2O$ .... 3.3 mM

$K_3Fe(CN)_6$ .... 3.3 mM

When required, add X-gal to a final concentration of 0.2% (from a 2% stock dissolved in dimethylsulfoxide and stored at $-20°C$)

Solution I

Glucose .... 50 mM

Tris Cl .... 25 mM

EDTA .... 10 mM

Adjust to pH 8.0

Solution II

NaOH 0.2N

SDS 1%

Solution III

Potassium acetate, 5 M .... 60 ml
Glacial acetic acid .... 11.5 ml

Dilute to 100 ml with UPW (28.5 ml).

TE

Tris HCl ...... 10 mM
EDTA .... 1 mM
Adjust to ...... pH 8.0

Sterilize by autoclaving

3. SAFETY PRECAUTIONS

Safety precautions for tissue culture and recombinant DNA procedures should be followed as recommended by local and international authorities. These are described in: "Guidelines for Small Scale Genetic Manipulation Work [1995]," available from the Genetic Manipulation Advisory Committee, GPO Box 2183, Canberra ACT 2601, Australia; and "Laboratory Biosafety Manual [1993]," available from the World Health Organisation, 1211 Geneva 27, Switzerland. In the United States, consult "Laboratory Safety Monograph: A Supplement to NIH Guidelines for Recombinant DNA Research [1978], National Institutes of Health, Bethesda, MD, USA, and in the UK, "A Guide to the Genetically Modified Organisms (Contained Use) Regulations 1992" HMSO Publication Centre, PO Box 276, London SW8 5DT, England.

4. DNA PREPARATION

Supercoiled plasmid DNA is transfected more efficiently than linear or open circular DNA [Chen and Okayama, 1987; Sorscher and Cordeiro-Stone, 1994]. Mammalian cells are very sensitive to toxins, and the purity of the plasmid DNA is critical to obtain consistent and efficient transfection efficiencies. Furthermore, because the amount of the DNA added to the precipitation cocktail has a profound affect on the nature of the precipitate formed and the efficiency of transfection, purity of the DNA solution for accurate spectrophotometric determination of a plasmid preparation is essential. Alkaline lysis of plasmid-bearing *E. coli* [Sambrook et al., 1989; Ausubel et al., 1995] followed by purification over anion-exchange columns (e.g., Qiagen) or CsCl equilibrium density gradient centrifugation yield plasmid DNA suitable for transfection. Qiagen columns are superior to one round of CsCl gradient centrifugation, but comparable to two rounds of CsCl treatment [Ehlert et al., 1993; Weber et al., 1995]. We have found that one round of CsCl gradient centrifugation followed by phenol-chloroform extraction and ethanol precipitation also yields DNA suitable for high-efficiency transfection.

When comparisons are to be made between the expression levels conferred by different plasmids, as when comparing various reporter gene constructs, the same purification method should be strictly adhered to for all plasmids. Given the adverse influence of contaminating lipopolysaccharides on transfection efficiencies [Weber et al., 1995], it is advisable to use the same bacterial host and media for the different plasmid preparations. Furthermore, multiple independent preparations of the same plasmid should be made and transfected to check for transfection consistency. The plasmid DNA should be prepared in sterile $0.1\times$ TE and treated aseptically.

4.1. Alkaline Lysis/CsCl Gradient Purification of Plasmid DNA

Protocol 4.1

Reagents and Materials
Aseptically prepared
- Plasmid-bearing *E. coli* (e.g., DH5α or XL1-Blue)
- Terrific Broth supplemented with the appropriate antibiotic
- TE

Nonsterile
- Solution I
- Solution II
- Solution III
- CsCl powder, 2.7 g, preweighed
- Ethidium bromide, 10 mg/ml 255 μl pre-measured
- 1-Butanol
- Isopropanol
- Ethanol, 70%
- Phenol, saturated solution in Tris HCl pH 8.0
- Phenol/chloroform 50:50
- Chloroform
- Sodium acetate, 3 M, pH 5.2
- Ethanol, absolute

Sterile equipment
- Centrifuge tubes, 50 ml
- Syringe and 18G needle

Nonsterile equipment
- Orbital shaker
- Centrifuge with Beckman JA-14 rotor or equivalent
- Beckman TL 100.3 rotor or equivalent
- Ultracentrifugation tubes, Beckman Quick-Seal No. 349621 or equivalent
- Corex tubes
- Microcentrifuge

Protocol
(a) Grow a 100 to 200-ml culture of plasmid-bearing *E. coli* (e.g., DH5α or XL1-Blue) in Terrific Broth supplemented with the appropriate antibiotic overnight, at 37°C in an orbital shaker.

(b) Centrifuge the cells in a Beckman JA-14 rotor at 5000 rpm (3800 g), 4°C for 5 min.

(c) Discard the supernatant and resuspend the bacterial pellet in 10 ml of Solution I.

(d) Add 20 ml of freshly prepared Solution II. Mix gently by inversion and let stand at room temperature for 10 min.

(e) Add 15 ml of ice-cold Solution III. Mix thoroughly by swirling and store on ice for 10 min.

(f) Centrifuge in a Beckman JA-14 rotor at 10,000 rpm (15,300 $\times$ g), 4°C for 10 min.

(g) Avoiding any flocculent white precipitate, collect 2 $\times$ 20 ml of supernatant into 2 $\times$ 50 ml centrifuge tubes. Add 12 ml of isopropanol to each tube and let stand at room temperature for 10 min.

(h) Pellet the plasmid DNA in a JA-17 rotor at 12,000 rpm (19,800 g), 20°C for 15 min.

(i) Discard the supernatant and carefully wash the pellet twice in 10 ml of 70% ethanol.

(j) Remove/drain all traces of ethanol. Briefly allow the ethanol to evaporate from the pellet. To avoid difficulties with resuspension avoid overdrying the pellet.

(k) Resuspend the pellet in 2.55 ml of TE.

(l) Add 2.7 g of CsCl and 255 μl ethidium bromide.

(m) Transfer all of the solution to an ultracentrifugation tube and centrifuge in a Beckman TL 100.3 rotor for either 5 h at 100,000 rpm (541,000 g) followed by 1 h at 85,000 rpm (346,000 g), or for 16 h at 85,000 rpm (346,000 g).

(n) Collect the closed circular plasmid DNA from the lower of the two upper bands with an 18G needle and syringe.

(o) In a microcentrifuge tube, add an equal volume of 1-butanol and mix to extract the ethidium bromide from the aqueous phase.

Briefly spin in a microcentrifuge to separate the two phases and collect the lower aqueous phase. Repeat this extraction procedure until all signs of ethidium bromide have been removed, as indicated by color disappearance.

(p) In Corex tubes, add three volumes of TE to the clear aqueous phase, followed by two final volumes of absolute ethanol. Allow the DNA to precipitate at room temperature for 15 min.

(q) Centrifuge in a Beckman JA-17 rotor at 9000 rpm (11,200 g) for 20 min.

(r) Carefully rinse the DNA pellet twice with 70% ethanol.

(s) Briefly dry the DNA pellet and resuspend in 300–600 μl of TE.

(t) Extract the DNA solution with an equal volume of phenol (pre-equilibrated with Tris HCl pH 8.0), and collect the upper aqueous phase after centrifugation in a microcentrifuge (maximum speed) for 3 min. Repeat the extraction procedure with equal volumes of phenol/chloroform, followed by chloroform.

(u) Precipitate the DNA by adding 0.1 volume of 3 M sodium acetate pH 5.2 and two final volumes of absolute ethanol. Microcentrifuge (maximum speed) for 15 min and carefully rinse the pellet twice with 1 ml of 70% ethanol. The final rinse should be performed aseptically in a laminar flow hood to avoid contamination.

(v) Briefly dry and resuspend the pellet in sterile 0.1× TE.

(w) Calculate the DNA concentration spectrophotometrically (1 OD_{260} unit corresponds to 50 μg/ml DNA).

5. CELL CULTURE

The human hepatoma cell lines, HuH7 [Nakabayashi et al., 1982] and HepG2 [Aden et al., 1979], the SV40-transformed African green monkey kidney cell line, COS [Gluzman, 1981], and the human cervical carcinoma cell line, HeLa [Scherer et al., 1953] are maintained as monolayer cultures in Dulbecco's modified Eagle's medium (DMEM) supplemented with 10% heat-inactivated fetal calf serum (FCS), penicillin (50 IU/ml), and streptomycin (50 μg/ml) and containing glutamine (2 mM) in a humidified atmosphere of 5% CO_2 at 36°C. On the day prior to transfection, HuH7, HepG2 or COS (0.7×10^5 cells/2-cm^2 well), or HeLa (0.3×10^5 cells/2-cm^2 well) cell lines are plated in 1 ml of DMEM/10% FCS per 2-cm^2 well in 4-well tissue culture plates (Nunc). The plating density may be varied according to cell type, expression level, and sensitivity of the assay. Generally it is chosen so as to achieve confluence at the time of assay.

6. CALCIUM PHOSPHATE TRANSFECTION

A common problem encountered by investigators using the calcium phosphate method to transiently transfect cell lines is the ability to obtain consistent efficiencies on a day-to-day basis. Since the conception of this method by Graham and Van der Eb [1973], it has been known that the nature and concentration of the DNA, the pH of the buffer in which the precipitate is formed, and the time that the cells are exposed to the precipitate significantly influence the efficiency of transfection [Loyter et al., 1982; Chen and Okayama, 1987; Pasco and Fagan, 1989; Sambrook et al., 1989; O'Mahoney et al., 1994; Ausubel et al., 1995; Jordan et al., 1996]. Most investigators follow one of two methods: the basic protocol involving the formation of a DNA precipitate with a HEPES-buffered solution prior to its addition over the cells, and a modified system employing a BES-buffered solution to form a DNA precipitate gradually in the supernatant medium during incubation [Chen and Okayama, 1987; Sambrook et al., 1989; Ausubel et al.,1995].

The method described below is a modification of the second procedure of Chen and Okayama [1987]. It was developed for the transient transfection of HepG2 and HuH7 hepatoma cell lines with luciferase reporter gene constructs [O'Mahoney and Adams, 1994], although it is applicable to other adherent cell lines, including COS and HeLa. In addition to the influences of DNA concentration and pH, we find that the standing time of the DNA-buffer solution prior to its addition to the cells is a parameter that must be strictly monitored to achieve the desired precipitate for maximum transfection efficiency. This finding has been confirmed by others optimizing the transfection of human embryo kidney HEK-293 cells, and Chinese hamster ovary cells, CHO [Jordan et al., 1996]. As such, on any given day, it is recommended that a few preliminary trial transfections be performed using two $2\times$ BBS preparations (e.g., pH 6.96 and 7.02), with different standing times (e.g., 0.5, 1.0, 1.5 min). The pH/standing time combination that results in the formation of a fine precipitate over the cells within 20 min is used for subsequent transfections. If a coarse precipitate forms in this time, either use a lower pH or decrease the standing time. It is advisable to transfect at least one plate with a positive control reporter gene plasmid that is under the control of a ubiquitous promoter. Such plasmids are available commercially: for example, pSV-β-gal (Promega), which expresses the *E. Coli* Lac Z gene under the control of the SV40 promoter and enhancer regions. This allows the transfection efficiency on any given day to be conveniently assessed by *in situ* staining for β-gal activity (see Protocol 7.1) prior to assaying other transfections.

Protocol 6

Reagents and Materials

- DNA solution: 2 μg plasmid/ml culture medium, 0.25 M $CaCl_2$ prepared from sterile stock solutions of plasmid DNA, and 2.5 M $CaCl_2$
- 2$\times$ BBS pH 6.96 or 7.02, pre-equilibrated to room temperature
- DMEM/10% FCS, or PBSA

Protocol

(a) For each volume of cell culture medium prepare 0.05 volume of DNA solution. Allow the solution to equilibrate to room temperature.

(b) Noting the time, quickly add and mix 0.05 volume of 2$\times$ BBS pH 6.96 or 7.02, and leave at room temperature for the predetermined standing time described above. At pH 7.02, the optimal time range is narrow (typically 0.5 or 1.0 min), while at pH 6.96 the times are longer.

(c) Once the standing time has elapsed, immediately add the solution dropwise to the cell culture while mixing (minimize the time that the cultures are out of the incubator, as CO_2/temperature equilibration influences the efficiency; see below).

(d) Incubate the cultures for 15–20 h at 36°C, 3% CO_2. This reduced CO_2 level was found to be optimal by Chen and Okayama [1987].

(e) Check the appearance of the cells and the nature of the precipitate under a phase contrast microscope with a 10$\times$ objective. A fine precipitate should be apparent on and between individual cells. If the precipitate is absent or coarse (often causing cell death), the transfections should be repeated using modified parameters (usually the pH or standing time).

(f) Remove the medium and rinse the cells with DMEM/10% FCS, or PBSA.

(g) Refeed the cells with DMEM/10% FCS and return to a 5% CO_2 incubator at 36°C for a further 22–27 h prior to assay. This incubation time should be determined empirically based upon the level of protein expression required for assay.

(h) Assay for expression.

7. REPORTER GENE ASSAYS

Reporter gene assay kits are available from a number of commercial suppliers (e.g., Promega, Clontech, Boehringer

Mannheim). Because of their sensitivity and ease of use, chemiluminescent detection kits for reporter genes are popular choices, with luminescence being detected in a luminometer or scintillation counter. Regardless of the detection method, it is recommended that a dilution series of pure enzyme be assayed to determine the instrument's linear range. Dilutions are normally made with the manufacturer's lysis buffer, supplemented with 1 mg/ml bovine serum albumin (BSA), to minimize the loss of enzyme activity due to absorption on to the assay tube's surface. In the case of β-gal, some kits may detect endogenous activity from the BSA itself. This activity may be avoided by heating the buffer at 48°C for 50 min.

When using scintillation counters to measure luminescence, most of it is detected within the low-energy range and can be measured on a tritium channel. It may be necessary with some instruments to calculate the square root of the counts minus background in order to achieve a linear relationship between the counts and the enzyme concentration. If significant background problems are encountered, they can often be attributed to static accumulation on gloves and tubes. Detailed protocols are usually supplied with the detection kits, and vary among manufacturers.

When selecting a kit, it is important to consider the sensitivity of the assay and the ease of use, with reference to the means of measuring the luminescence. Kits requiring the addition of only a single reagent to cell lysates offer a distinct advantage, not only with respect to convenience, but also by eliminating any variability associated with the time taken to add successive reagents with multiple reagent kits. For assays that generate stable luminescent signals lasting minutes rather than seconds it is possible to use scintillation counters and luminometers without automatic injection devices. Promega offers a sensitive, single-reagent luciferase assay that generates a signal with a half-life of ~10 min. As the signal is relatively stable for the first minute, there is sufficient time for mixing and analysis with a scintillation counter.

Packard Instruments supplies a luciferase assay kit that generates a luminescent signal with a half-life of several hours. With this method, it is possible to prepare numerous samples, which may be analyzed with instruments designed for 96-well plates. A chemiluminescent assay system from Clontech for secreted alkaline phosphatase involves measurements on culture medium aliquots; this method avoids cell disruption and enables numerous samples to be analyzed over a time course. This is particularly attractive for studies in which the reporter is being used to assess the cells' response to a particular treatment (e.g., hormones or transfected expression constructs). Cell lysis buffers that are compatible with multiple reporter assays are convenient for measuring co-transfected internal control reported genes. A dual luciferase reporter assay system, which enables the detection of two types of luciferase, has recently been released by Promega.

Promega also supply a β-gal assay kit containing the substrate ONPG (o-nitrophenyl-β-D-galactopyranoside), with the reaction monitored spectrophotometrically at a wavelength of 420 nm. These kits may be used to monitor transfection efficiency from positive control plasmids such as pSV-β-gal (Promega). Alternatively, if a β-gal-positive control plasmid is transfected, a visual indication of the percentage of cells transfected may be obtained by *in situ* staining. Following the transfection procedure described, a good correlation exists between the percentage of cells staining positive for β-gal and the activities of test reporter plasmids transfected in parallel.

7.1. β-Gal *In Situ* Staining

Protocol 7.1

Reagents and Materials
- PBSA
- Glutaraldehyde solution
- X-gal solution

Protocol
(a) Rinse the transfected cells twice with PBSA.
(b) Fix with glutaraldehyde solution for 15 min at room temperature.
(c) Remove the glutaraldehyde solution and replace it with X-gal solution.
(d) Incubate at 37°C for 1 h. Transfectants appear blue with varying intensities.

8. PARAMETERS THAT INFLUENCE PRECIPITATE FORMATION AND TRANSFECTION EFFICIENCY

The limitations of internal control plasmids in correcting for variations in transfection efficiency may be circumvented by controlling the parameters influencing calcium phosphate-DNA co-precipitate formation. The nature of the precipitate is directly related to the subsequent transfection efficiency, an abundance of a fine precipitate over the cells being optimal [Chen and Okayama, 1987; O'Mahoney and Adams, 1994; Jordan et al., 1996]. Such a precipitate exhibits Brownian motion and is superior to coarse, less abundant precipitates, which often cause cell death.

8.1. Interaction of pH, Standing Time, and DNA Concentration

Day-to-day variability in the nature of the precipitate and transfection efficiency can be controlled by a number of interacting parameters. As shown in Figs. 7.2 and 7.3, the combination of pH, plasmid concentration, and standing time have profound effects on transfection efficiency as revealed by reporter gene analysis. While the pH of the buffer, and the amount of DNA have always been considered to be important, the standing time (the length of time that the DNA/CaCl$_2$/2× BBS is left to stand at room temperature prior to its addition over the cells) is a parameter that is usually overlooked. Most procedures recommend a standing time of 10 to 20 min. It has recently been observed [O'Mahoney and Adams, 1994; Jordan et al., 1996] that precipitate formation occurs very rapidly, particularly with buffers in the higher pH range (e.g., pH 7.02). As a result, the formation of an abundant fine precipitate (optimal) as opposed to a less abundant coarse precipitate occurs within a small time frame. By strictly adhering to the predetermined optimal standing time on any given day, using the same reagents with a constant amount of DNA, consistent transfection efficiencies can be achieved. It is convenient to select a DNA concentration/pH combination such that the optimal standing time is minimized (e.g., 0.5 or 1.0 min). Following mixing of the DNA/CaCl$_2$/2× BBS solution, it can then be added almost immediately

Fig. 7.2. The pH and standing time influences the transfection efficiency. HuH7 cells cultured in 4-well plates were transfected with 2 μg of a luciferase reporter plasmid using 2× BBS at pH 7.02 or pH 6.96. Following the addition 2× BBS, different standing times were tested before the DNA preparation was added to the cells. Each well was assayed for luciferase activity (cpm $\times$ 10^{-3}) and the mean $\pm$ SD of four independent transfections for each combination is presented. Reproduced from O'Mahoney and Adams [1994], with permission from the publisher.

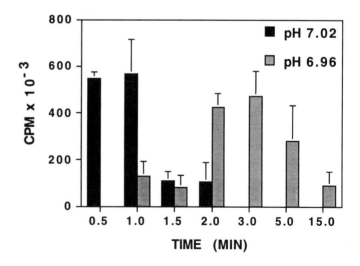

O'Mahoney and Adams

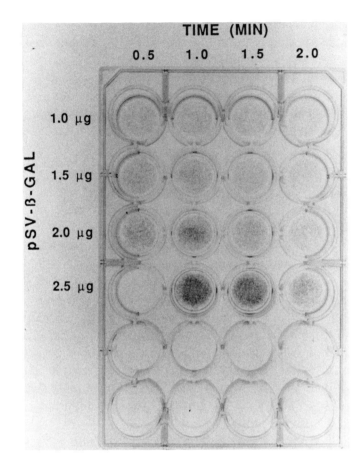

TIME (MIN)

| | 0.5 | 1.0 | 1.5 | 2.0 |

pSV-β-GAL

1.0 µg

1.5 µg

2.0 µg

2.5 µg

Fig. 7.3. The plasmid concentration and standing time influences the transfection efficiency. HuH7 cells were transfected with 1–2.5 µg of pSV-β-gal, with different standing times following the addition of 2× BBS at pH 7.02. The cells were stained *in situ* for β-gal. Reproduced from O'Mahoney and Adams [1994], with permission from the publisher.

to the cells without having to keep a record of the standing times for a backlog of several transfections. This is advantageous for performing large numbers of transfections, as frequently required for replicate reporter gene experiments.

8.2. Temperature and Culture Conditions

The temperature of the DNA/CaCl$_2$/2× BBS mix has a significant influence on precipitate formation. In our experience, the optimal standing times were often reduced on days when the temperature of the tissue culture room was abnormally high. Jordan and colleagues have accounted for this observation by showing that higher temperatures increase the rate of precipitate formation and that even small variations can affect its kinetics

[Jordan et al., 1996]. This highlights the importance of equilibrating the solutions to room temperature and of keeping them away from heat sources (e.g., Bunsen burner).

Allowing the culture medium to equilibrate to room temperature and atmospheric CO_2 prior to adding the DNA solution to the cells may be detrimental to the achievement of high efficiency transfection for a given set of parameters. As shown in Fig. 7.4, by allowing equilibration over a period of 40 min, the efficiency of transfection by two constructs is drastically reduced based on the assay of their relative reporter genes. This is an important consideration when using multiwell tissue culture plates. In general, it is advisable to limit the number of wells per plate to 4. Beyond this, other wells will have become partially or fully equilibrated by the time four transfections have been performed, and the efficiency of transfection begins to decline, contributing to unnecessarily high variations with replicate transfections. Furthermore, if large numbers of plates are to be accessed, it is wise to distribute them into two or more incubators and to alternate their access in order to allow the CO_2 levels and temperature to recover. Four-well plates are convenient for quadruplicate reporter gene transfections in the absence of co-transfected internal controls. Variation between the first and last well of a plate due to the equilibration effect can be minimized by working quickly, and by using the pipette tip to stir the culture medium while gradually expelling the DNA solution rather than swirling all four

Fig. 7.4. Media equilibration adversely effects the transfection efficiency. HuH7 cells cultured in 4-well plates were transfected with 1.98 μg of pGHR-GL2 (a luciferase reporter plasmid) + 20 ng of pCMV-β-gal in quadruplicate, using a plate of cells taken immediately from a 5% CO_2/36°C incubator (fresh plate) or a plate that had been equilibrated to atmosphere/room temperature for 40 min. Cell lysates were assayed for luciferase and β-gal activity. The mean and SD expressed are relative to those of a fresh plate (100%). Reproduced from O'Mahoney and Adams [1994], with permission from the publisher.

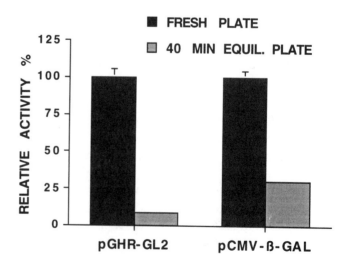

O'Mahoney and Adams

wells. Swirling the whole plate tends to accelerate the equilibration of all four wells.

9. VARIATIONS OF THE CALCIUM PHOSPHATE TRANSFECTION METHOD

The procedure described above is capable of achieving transfection efficiencies as high as 100% for HuH7 and HepG2 hepatoma cells, and 40% for HeLa and COS cells, when cultured in DMEM/10%FCS. A number of modifications to the procedure may be required for other cell types or media conditions. However, whatever the modifications, the parameters of DNA concentration, pH, standing time, and plate equilibration should remain as important considerations. Jordan and colleagues have demonstrated that such parameters are applicable to both transient and stable transfections [Jordan et al., 1996]. Pasco and Fagan [1989] identify a number of conditions that influence the transfection of primary hepatocytes by the method of Chen and Okayama [1987]. These include the presence of protein in the medium, the nature of the cell substratum, cell density, duration of exposure to the precipitate, and the age of the culture. Such conditions often need to be manipulated and thus optimized for certain cell types so as to maintain their phenotype.

Optimization of calcium phosphate transfection for bovine chromaffin cells and the African green monkey cell line, BSC-40, reveals that the type of buffer used for precipitate formation may also play a role in transfection efficiency [Wilson et al., 1995]. Interestingly, the optimal conditions employed by Wilson and colleagues include the absence of serum in the culture medium, contrasting with the protein requirements of Pasco and Fagan [1989], and the presence of FCS in the media of most reports. Calcium and phosphate concentrations also have an obvious influence on precipitate formation and may be adjusted accordingly [Wilson et al., 1995; Jordan et al., 1996]. Generally, however, it is more convenient to manipulate the standing time. Using the standard HEPES buffer procedure, the transfection of some cell types may be enhanced by a glycerol and/or DMSO shock [Ausubel et al., 1995; Wilson et al., 1995; Jordan et al., 1996]. This is not normally required with the modified BES-buffered procedure [Ausubel et al., 1995]; however, it could be a consideration for difficult cell types.

The procedure of Chen and Okayama [1987] recommends that, following the addition of the DNA to the cells, they be cultured at a lower CO_2 level (3%) while the precipitate forms over the cells. At this level, the medium is slightly basic (approximately pH 7.6) and may act to counter the effects of the acidic precipitate solution and lactic acid production by the cells. This step has been maintained in the above protocol, although its requirement may vary with cell type and media.

REFERENCES

Aden DP, Fogel A, Plotkin S, Damjanov I, Knowles BB (1979): Controlled synthesis of HBsAg in a differentiated human liver carcinoma-derived cell line. Nature 282: 615–616.

Alam J, Cook JL (1990): Reporter genes: applications to the study of mammalian gene transcription. Anal Biochem 188: 245–254.

Ausubel FM, Brent R, Kingston RE, Moore DD, Seidman JG, Smith JA, Struhl K (eds) (1995): "Current Protocols in Molecular Biology." New York: John Wiley & Sons.

Bergeron D, Barbeau B, Leger C, Rassart E (1995): Experimental bias in the evaluation of the cellular transient expression in DNA co-transfection experiments. Cell Mol Biol Res 41: 155–159.

Chen C, Okayama H (1987): High-efficiency transformation of mammalian cells by plasmid DNA. Mol Cell Biol 7: 2745–2752.

Ehlert F, Bierbaum P, Schorr J (1993): Importance of DNA quality for transfection efficiency. Biotechniques 14: 546.

Farr A, Roman A (1992): A pitfall of using a second plasmid to determine transfection efficiency. Nucleic Acids Res 20: 920.

Gluzman Y (1981): SV40-transformed simian cells support the replication of early SV40 mutants. Cell 23: 175–182.

Graham FL, van der Eb (1973): A new technique for the assay of infectivity of human adenovirus 5 DNA. Virology 52: 456–467.

Hirt B (1967): Selective extraction of polyoma DNA from infected mouse cell cultures. J Mol Biol 26: 365–369.

Jordan M, Schallhorn A, Wurm FM (1996): Transfecting mammalian cells: optimization of critical parameters affecting calcium phosphate precipitate formation. Nucleic Acids Res 24: 596–601.

Levitt LK, O'Mahoney JV, Brennan KJ, Joya JE, Zhu L, Wade RP, Hardeman EC (1995): The human troponin I slow promoter directs slow fiber-specific expression in transgenic mice. DNA Cell Biol 14: 599–607.

Loyter A, Scangos GA, Ruddle FH (1982): Mechanisms of DNA uptake by mammalian cells: fate of exogenously added DNA monitored by the use of fluorescent dyes. Proc Natl Acad Sci USA 79: 422–426.

Nakabayashi H, Taketa K, Miyano K, Yamane T, Sato J (1982): Growth of human hepatoma cell lines with differentiated functions in chemically defined medium. Cancer Res 42: 3858–3863.

O'Mahoney JV, Adams TE (1994): Optimization of experimental variables influencing reporter gene expression in hepatoma cells following calcium phosphate transfection. DNA Cell Biol 13: 1227–1232.

Pasco DS, Fagan JB (1989): Efficient DNA-mediated gene transfer into primary cultures of adult rat hepatocytes. DNA 8: 535–541.

Ravid K, Takefumi D, Beeler DL, Kluter DJ, Rosenberg RD (1991): Transcriptional regulation of the rat platelet factor 4 gene: interaction between an enhancer/silencer domain and the GATA site. Mol Cell Biol 11: 6116–6127.

Sambrook J, Fritsch EF, Maniatis T (eds) (1989): "Molecular Cloning: A Laboratory Manual," 2nd ed. New York: Cold Spring Harbor Laboratory Press.

Scherer WF, Syverton JT, Gey GO (1953): Studies on the propagation *in vitro* of poliomyelitis viruses. J Exp Med 97: 695–710.

Sorscher DH, Cordeiro-Stone M (1994): Inhibition of reporter gene expression in mammalian cells. Effects of carcinogen lesions in DNA. Carcinogenesis 15: 1093–1096.

Weber M, Moller K, Welzeck M, Schorr J (1995): Effects of lipopolysaccharide on transfection efficiency in eukaryotic cells. Biotechniques 19: 930–940.

Wilson SP, Liu F, Wilson RE, Housley PR (1995): Optimization of calcium phosphate transfection for bovine chromaffin cells: relationship to calcium phosphate precipitate formation. Anal Biochem 226: 212–220.

APPENDIX: MATERIALS AND SUPPLIERS

Material	Supplier
BES [N,N-bis(2-hydroxyethyl)-2-aminoethanesulfonic acid]:	Sigma
BSA	Boehringer Mannheim
DMSO	Sigma
Dulbecco's modified Eagle's medium	Gibco BRL
Fetal calf serum	Gibco BRL
Glutaraldehyde	Sigma
Penicillin	Gibco BRL
pSV-β-gal	Promega
Qiagen columns	Qiagen
Reporter gene detection kits and plasmids	Boehringer Mannheim, Clontech, Promega
Streptomycin	Gibco BRL
Tissue culture plates, 4-well	Nunc
Tryptone and yeast extract	Difco
X-gal	Sigma

8

Strontium Phosphate Transfection

Roger R. Reddel, Curtis C. Harris, and Douglas E. Brash

*The Children's Medical Research Institute, 214 Hawkesbury Rd,
Westmead, Sydney, NSW 2145 Australia (R.R.R.); Laboratory of Human
Carcinogenesis, National Cancer Institute, National Institutes of Health,
37 Convent Drive, Building 37, Rm 2C01, Bethesda, Maryland 20892-
4255 (C.C.H.); Department of Therapeutic Radiology, Yale School of
Medicine, 15 York St., New Haven, Connecticut 06520-8040 (D.E.B.)*

DNA Transfer to Cultured Cells, Edited by Katya Ravid and R. Ian Freshney.
ISBN 0-471-16572-7   © 1998 Wiley-Liss, Inc.

I. INTRODUCTION AND BACKGROUND

Transfection of mammalian cells via DNA-strontium phosphate co-precipitation [Brash et al., 1987] is essentially a variant of the calcium phosphate method. As the latter procedure has been described in detail in the preceding two chapters, its strontium counterpart will be described very briefly here.

Although calcium phosphate transfection [Graham and Van der Eb, 1973; Wigler et al., 1979] is a versatile and widely used procedure, there are some cell types for which it proved to be unsuitable. For example, it was found that normal human bronchial epithelial cells (NHBE) were lysed by calcium phosphate precipitates, and ionic calcium in the presence of serum induces these and other cells (including rat, mouse, and human keratinocytes) to undergo squamous differentiation [Hennings et al., 1980; Praeger and Gilchrest, 1986].

Strontium phosphate transfection has been used successfully to transfect NIH 3T3 cells as well as the following types of normal human cells: NHBE, esophageal epithelial, fetal prostatic epithelial, normal mesothelial, mastlike cells, hepatocytes, and fibroblasts [Brash et al., 1987; Reddel et al., 1988; Ke et al., 1989; Kaighn et al., 1989; Lechner et al., 1989; Stoner et al., 1991; Reddel et al., 1991; Townsend et al., 1993; De Silva and Reddel 1993]. It has also been used to transfect immortalized cells and tumor cell lines [Reddel et al., 1989; Noble et al., 1992; De Silva et al., 1994; Reddel et al., 1995]. Strontium lies directly below calcium on the periodic chart and makes similar phosphate crystals.

2. PREPARATION OF REAGENTS

2.1. Cell Culture HEPES Buffered Saline (HBS)

HEPES	4.76 g
NaCl	7.07 g
KCl	0.20 g
Glucose	1.70 g
Na_2HPO_4	1.022 g
Phenol red solution, 0.5% w/v	0.25 ml
UPW to	900 ml

Adjust the pH to 7.5 with 1.0 M NaOH.
Make the final volume to 1 liter with UPW.

Filter-sterilize the solution (0.2-μm filter) and store at room temperature.

2.2. Stock Solutions for 20× Transfection-HBS

(a) Add 1.88 g of $Na_2HPO_4.7H_2O$ to 500 ml of distilled water.
(b) Add 10.8 g of dextrose to 500 ml of distilled water.

(c) Add 80 g of NaCl to 500 ml of distilled water.
(d) Add 3.72 g of KCl to 500 ml of distilled water.
(e) Add 47.6 g of HEPES to 500 ml of distilled water.

Store each of these five stock solutions at $-20°C$.

2.3. 2× Transfection-HBS

Combine 10 ml of each of the 20× stock solutions (above). Adjust the pH to the optimal level (see below). Make the total volume up to 100 ml with UPW. Double filter (see Section 2.7 below) and store at 4°C.

2.4. Optimizing pH of 2× Transfection-HBS Solution

The transfection efficiency is critically dependent on the quality of the precipitate, which in turn is dependent on the pH of the DNA–strontium phosphate mix. The optimal pH of the 2× transfection-HBS solution is determined by the characteristics of the serum-free growth medium and the CO_2 concentration of the incubator. Use the same batch of medium and the same incubator throughout, and test 2× transfection-HBS at a range of pH values (e.g., at intervals of 0.05 pH units from 7.20 to 7.80) using the protocol described below. The optimum pH may be assessed either on the expression of a reporter gene (e.g., see Chapter 7) or on the quality of the precipitate (Fig. 8.1). Aim to produce a copious fine dustlike precipitate that carpets the cells and adheres to them and the surface of the dish.

2.5. Glycerol/1× Transfection-HBS

Glycerol .... 15 g
2× transfection-HBS ......   50 ml
UPW to ....   100 ml

Double-filter (see Section 2.7 below) and store at 4°C.

2.6. SrCl₂ Solution

Make 100 ml of a 2 M solution of $SrCl_2$ in UPW. Double-filter (see Section 2.7) and store at 4°C. It is important to use ultrapure strontium, as the main impurity in commercially available strontium is calcium.

2.7. Filtration of Reagents

The 2× transfection-HBS, glycerol/1× transfection-HBS, water, and $SrCl_2$ solutions used in strontium phosphate transfections are each filtered through two 0.2-μm Nalgene filtration units (i.e., double-filtered). The brand of filtration unit used has a substantial effect on transfection efficiency. Rinse each filter by discarding the initial 4–5 ml of filtrate, in order to avoid aggregation of the precipitate.

Fig. 8.1. Phase contrast photomicrographs of NHBE cells during strontium phosphate transfection. Reporter gene assays showed that the copious fine adherent precipitate formed at pH 7.3 was optimal (**A**). The reticular precipitate at pH 7.4 (**B**) was less effective, and the coarse aggregates formed at pH 7.5 (**C**) produced excessive cytotoxicity. The pH needs to be optimized for each combination of cell culture medium and CO_2 incubator as described in the text.

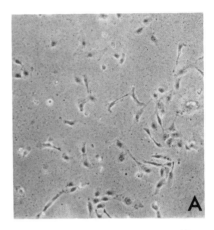

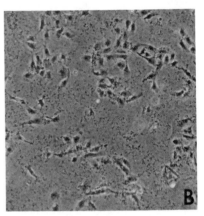

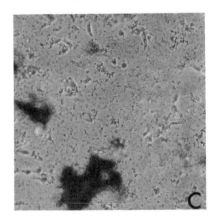

Reddel et al.

2.8. Plasmid DNA

DNA may be prepared for transfection by two CsCl ultracentrifugation steps [Sambrook et al., 1989] or by Qiagen column chromatography.

2.9. Serum-free Growth Medium

For transfection, use the cells' usual growth medium if it is serum-free. If not, prepare serum-free medium by omitting serum from the usual formulation. The medium without any additives is referred to in the method as "basal medium."

3. METHOD OF TRANSFECTION

Protocol 3

Reagents and Materials

Aseptically prepared
- Cells for transfection
- Growth medium, serum-free (as appropriate for cell type)
- Growth medium with serum as appropriate
- HBS
- Polypropylene tubes
- 2× transfection-HBS
- UPW, double-filtered
- Plasmid DNA
- SrCl$_2$, 2 M (ultrapure, calcium-free)
- Glycerol/1× transfection-HBS

Nonsterile
- Heating block

Protocol 3

(a) On the afternoon before transfection, harvest and count cells, and seed at a concentration of 0.5–1 × 10^6 cells per 100-mm dish. Return cells to the humidified 37°C CO$_2$ incubator overnight.

(b) On the day of transfection, pre-equilibrate the serum-free growth medium in the 37°C CO$_2$ incubator for at least 1 h.

(c) Remove the medium and wash the cells three times with 10 ml of HBS.

(d) Add 10 ml of pre-equilibrated serum-free growth medium to the cells and return to the incubator for approximately 1 h (maximum 3 h).

(e) Pre-warm polypropylene tubes (one tube for each dish of cells) in a 37°C heating block during the 1-h incubation.

(f) Add the following reagents in the specified order to each polypropylene tube:

     (i)   2× transfection-HBS, 500 μl

     (ii)   Double-filtered UPW, 440 μl minus the volume of plasmid DNA (x). [see iii below]

     (iii) Plasmid DNA, 10–20 μg per dish (x μl)

     (iv) 2 M $SrCl_2$, 62 μl

(g) Gently agitate each polypropylene tube and pipette up and down once in a sterile 1-ml pipette to mix.

(h) Return each tube to the 37°C heating block and incubate for exactly 30 s.

(i) Draw the DNA-strontium-phosphate mix from the polypropylene tube using a 1-ml pipette and add dropwise to the medium in a dish of cells.

(j) Rock each dish gently to mix each drop as it is added and return to the incubator for approximately 2 h.

(k) After 1–2 h use a microscope with a 10× or 20× objective to observe whether a fine dust-like precipitate has settled on the cells.

(l) Remove and discard the medium and incubate the cells with 3 ml glycerol/1× transfection-HBS solution for 30 s at room temperature.

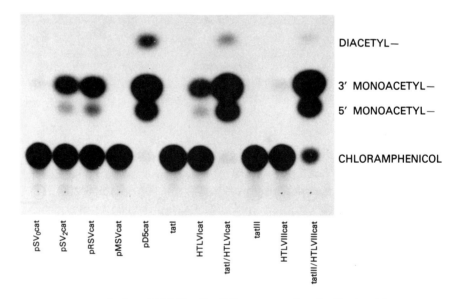

Fig. 8.2. Transfection of NHBE cells. Cells were transfected with plasmid constructs containing the chloramphenicol acetyltransferase (CAT) gene linked to various promoter/enhancers. After 2 days CAT activity was assayed by a standard protocol. The CAT gene was linked to a promoterless SV40 enhancer (pSV₀), SV40 early promoter/enhancer (pSV₂), Rous sarcoma virus long terminal repeat (LTR; pRSV), Moloney sarcoma virus LTR (pMSV), adenovirus major late promoter (pD5), HTLVI LTR, or the HTLVIII (HIV1) LTR. Plasmids encoding the HTLVI or HIV1 tat proteins (tatI and tatIII, respectively) were included in some transfections.

Reddel et al.

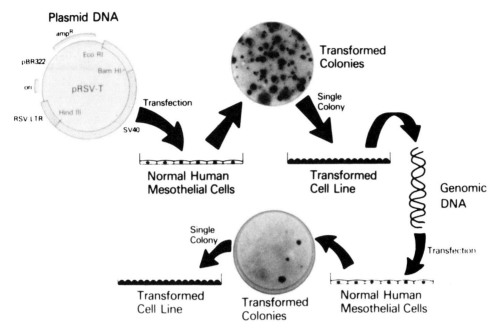

Fig. 8.3. Transfection of normal human mesothelial cells with plasmid and human genomic DNA. Cells were transfected via strontium phosphate co-precipitation with a plasmid, pRSV-T, encoding the SV40 early region linked to the Rous sarcoma virus LTR. A colony of transformed cells was isolated and passaged as the MeT-5A cell strain (Ke et al., 1989). Genomic DNA was isolated from MeT-5A cells by standard methods, purified by two rounds of ultracentrifugation through CsCl, and transfected by the strontium phosphate method into a culture of normal mesothelial cells obtained from a different individual. Transformed colonies were isolated and passaged to obtain cell strains that expressed SV40 T antigens (not shown).

(m) Immediately remove and discard the glycerol solution and wash the cells 3 times with basal medium at room temperature.

(n) Feed the cells with 10 ml of growth medium and return to the incubator.

An example of the use of strontium phosphate transfection of NHBE cells is shown in Fig. 8.2.

4. VARIATIONS ON THE METHOD

A variety of techniques for producing precipitates have been developed for calcium phosphate transfection, and presumably many or all of these would also be successful if strontium were substituted for calcium. The above protocol is based on the addendum to Brash et al., [1987]; an alternative procedure is described within the text of the same paper.

Strontium phosphate co-precipitation has also been used to transfect primary human cultures with high molecular weight genomic DNA purified by two rounds of CsCl ultracentrifugation (Fig. 8.3).

ACKNOWLEDGMENTS

The authors thank Myra Quanrud, Kirsten Alcorn, James Laird, and Trevor Johnson for assistance with the transfections.

REFERENCES

Brash DE, Reddel RR, Quanrud Y, Ke Y, Farrell MP, Harris CC (1987): Strontium phosphate transfection of human cells in primary culture: stable expression of the simian virus 40 large-T-antigen gene in primary human bronchial epithelial cells. Mol Cell Biol 7: 2031–2034.

De Silva R, Reddel RR (1993): Similar simian virus 40-induced immortalization frequency of fibroblasts and epithelial cells from human large airways. Cell Mol Biol Res 39: 101–110.

De Silva RA, Englezou A, Schevzov G, Gerwin BI, Harris CC, Gunning P, Reddel RR (1994): Induction of anchorage independent growth and serum resistance in immortalized human bronchial epithelial cells by alteration of the cytoskeleton. Cell Mol Biol Res 40: 323–335.

Graham FL, Van der Eb AJ (1973): A new technique for the assay of infectivity of human adenovirus 5 DNA. Virology 52: 456–467.

Hennings H, Michael D, Cheng C, Steinert P, Holbrook K, Yuspa SH (1980): Calcium regulation of growth and differentiation of mouse epidermal cells in culture. Cell 19: 245–254.

Kaighn ME, Reddel RR, Lechner JF, Peehl DM, Camalier RF, Brash DE, Saffiotti U, Harris CC (1989): Transformation of human neonatal prostate epithelial cells by strontium phosphate transfection with a plasmid containing SV40 early region genes. Cancer Res. 49: 3050–3056.

Ke Y, Reddel RR, Gerwin BI, Reddel HK, Somers ANA, McMenamin MG, LaVeck MA, Stahel RA, Lechner JF, Harris CC (1989): Establishment of a human in vitro mesothelial cell model system for investigating mechanisms of asbestos-induced mesothelioma. Am J Pathol 134: 979–991.

Lechner JF, Cole KE, Reddel RR, Anderson L, Harris CC (1989): Replicative cultures of adult human and rhesus monkey liver epithelial cells. Cancer Detec Prev 14: 239–244.

Noble JR, Willetts KE, Mercer WE, Reddel RR (1992): Effects of exogenous wild-type p53 on a human lung carcinoma cell line with endogenous wild-type p53. Exp Cell Res 203: 297–304.

Praeger FC, Gilchrest BA (1986): Strontium stimulates human keratinocyte proliferation. Clin Res 34: 775A.

Reddel RR, Ke Y, Gerwin BI, McMenamin MG, Lechner JF, Su RT, Brash DE, Park J-B, Rhim JS, Harris CC (1988): Transformation of human bronchial epithelial cells by infection with SV40 or adenovirus-12 SV40 hybrid virus, or transfection via strontium phosphate coprecipitation with a plasmid containing SV40 early region genes. Cancer Res 48: 1904–1909.

Reddel RR, Malan-Shibley L, Gerwin BI, Metcalf RA, Harris CC (1989): Tumorigenicity of human mesothelial cell line transfected with EJ-ras oncogene. J Natl Cancer Inst 81: 945–948.

Reddel RR, Hsu I-C, Mass MJ, Hukku B, Gerwin BI, Salghetti SE, Somers ANA, Galati AJ, Gunning WT III, Harris CC, Stoner GD (1991): A human bronchial epithelial cell strain with unusual *in vitro* growth potential which undergoes neoplastic transformation after SV40 T antigen gene transfection. Int J Cancer 48: 764–773.

Reddel RR, De Silva R, Duncan EL, Rogan EM, Whitaker NJ, Zahra DG, Ke Y, McMenamin MG, Gerwin BI, Harris CC (1995): SV40-induced immortalization and *ras*-transformation of human bronchial epithelial cells. Int J Cancer 61: 199–205.

Sambrook J, Fritsch EF, Maniatis T (1989): "Molecular Cloning. A Laboratory Manual." New York: Cold Spring Harbor Laboratory Press.

Stoner GD, Kaighn ME, Reddel RR, Resau JH, Bowman D, Naito Z, Matsukura N, You M, Galati AJ, Harris CC (1991): Establishment and characterization of SV40 T-antigen immortalized human esophageal epithelial cells. Cancer Res 51: 365–371.

Townsend M, Macpherson J, Krilis S, Reddel R, Symonds G (1993): Immortalization and characterization of human cell lines with mast cell and monocytic properties. Br J Haematol 85: 452–461.

Wigler M, Pellicer A, Silverstein S, Axel R, Urlaub G, Chasin L (1979): DNA-mediated transfer of the adenine phosphoribosyltransferase locus into mammalian cells. Proc Natl Acad Sci USA 76: 1373–1376.

APPENDIX: LIST OF SUPPLIERS

Materials	Catalogue Number	Suppliers
Dextrose	17-0340-01	Pharmacia LKB
Disodium hydrogen phosphate	10249	BDH/Merck
Glucose	G7520	Sigma Chemical Co.
Glycerol (analytical grade)	242	Ajax Chemicals
HEPES	H9136	Sigma Chemical Co.
Nalgene filtration units, 150 ml 0.2 μm	125-0020	Nalge Company
Phenol red	20090	BDH/Merck
Polypropylene tubes	2059	Becton Dickinson
Potassium chloride	383	Ajax Chemicals
Qiagen Plasmid Mega Kit	12181	Qiagen Inc.
Sodium chloride	10241	BDH/Merck
Sodium hydroxide	10252	BDH/Merck
Strontium chloride	14029	BDH/Merck

9

In Vitro Tissue Transfection by Calcium Phosphate

M. F. James, C. B. Rich, V. Trinkaus-Randall, and J. A. Foster

Department of Biochemistry, Boston University School of Medicine, 80 East Concord Street, Boston, Massachusetts 02118

DNA Transfer to Cultured Cells, Edited by Katya Ravid and R. Ian Freshney.
ISBN 0-471-16572-7   © 1998 Wiley-Liss, Inc.

1. INTRODUCTION TO TISSUE TRANSFECTIONS

The transient expression of foreign DNA in intact tissue has proven to be a valuable tool for studying gene regulation. Although the transient transfection of reporter constructs in isolated cell cultures has been an effective technique for revealing potential mechanisms of gene control, the environment of a homogeneous cell population represents a less than accurate model of the *in vivo* system. Gene expression in complex tissues and organs is dependent on regulatory signals that require specific and complex cell–cell and cell–matrix interactions that cannot be reproduced in primary cell culture [Arnold et al., 1994]. Lung tissue, for example, encompasses a heterogeneous cell population consisting of over 40 different cell types [Kuhn, 1978]. Cellular differentiation and the maintenance of phenotype require the proper temporal and spatial developmental cues, as directed by heterotypic cell contact, the expression of specific genes, and the endocrine and paracrine delivery of key modulators [Moscona and Linser, 1983; Arnold et al., 1994].

2. METHODS USED TO TRANSFECT TISSUE

Several gene transfer methodologies are currently being used in intact tissue to study gene expression more accurately within the *in vivo* environment. Pu and Young [1990] have compared the efficiency of three such techniques in order to study the transcriptional mechanisms regulating the glutamine synthetase (Glns) gene in intact chick retinal explant cultures. Electroporation, liposome, and calcium phosphate coprecipitation methods of gene transfer were examined by transfecting *cat* or *lacZ* reporter constructs under the transcriptional control of viral promoters or the chicken Glns promoter. These studies show the highest recovered activity levels are obtained using extracts from the electroporated explants. The calcium phosphate technique resulted in moderate activity, while undetectable levels resulted from lipofection. These approaches have revealed that the Glns gene is glucocorticoid-inducible and that mediating *cis*-acting elements localize to an 1.2-kb upstream region. In addition, the Glns gene has been shown to be responsive to glucocorticoid induction only when neuronal and Muller cells remain in contact [Moscona and Linser, 1983;

Linser and Moscona, 1979; Linser and Perkins, 1987]. Other researchers [Zhang and Young, 1991; Ben-Dror et al., 1993; Berko-Flint et al., 1994] also report successful explant transfection results using similar methods of transfection.

Arnold et al. [1994] have demonstrated a method that combines particle bombardment transfection procedures and organotypic slice culture techniques to study transcriptional control in the developing nervous system. This system is advantageous because nearly 100% co-transfection efficiency of multiple DNAs can be achieved. That is, every positively transfected cell will express two constructs, allowing each cell a perfectly designed internal control while providing a wild-type environment for the study of gene function.

Several laboratories have successfully transfected tissue *in vitro* by applying plasmid DNA complexed to cationic liposomes [Fisher et al., 1996; Takeshita et al., 1994]. Strategies for arterial gene transfer utilizing a rabbit aortic model system have been developed for the potential study of a variety of cardiovascular disorders [Takeshita et al., 1994]. Specifically, this study revealed that the expression of liposome-mediated arterial gene transfer may be increased in the presence of ongoing cellular proliferation.

3. ADVANTAGES OF TRANSIENT TISSUE TRANSFECTIONS

Dissection of regulatory sequences that are potential targets of known transcriptional regulatory proteins often requires the transient transfection of viral vectors or the use of transgenic mice [Arnold et al., 1994]. With such methods, analysis of the expression patterns of several different fusion gene constructs in multiple transgenic lines or reconstructed viruses can often take many months or years to complete. *In vitro* tissue transfection and processing of tissue takes approximately 1 week. Although this method is not applicable to many of the goals addressed by transgenic approaches, such as gene knock-outs and general investigations into over-expression of specific transgenes, it does offer a method to determine the functionality of specific promoter sequences within a tissue environment.

We have incorporated a new approach to determine the function of regulatory *cis*-acting sequences *in vivo* that allows both the quantitation and localization of reporter activities. This method identifies gene reporter expression activity in a whole-tissue environment by immunohistochemical analysis. In addition, it affords a dynamic system dictated by *in vivo* factors in which to study gene control, and thereby avoids the undifferentiated state and potential phenotypic drift associated with cultured cells. This technique also allows transfection of all cell types with relatively high efficiency as demonstrated by immunohistochemical analysis of the co-transfected β-gal expression vector, thereby permitting the identification of promoter sequences that are capable of reproducing the same expression patterns as the endogenous gene.

4. REAGENTS AND SUPPLIES

4.1. Sterile Reagents for Tissue Transfection

Complete medium with 10% serum (DMEM/10FB)
Dulbecco's modified Eagle's medium (DMEM)
10% Fetal bovine serum (FBS)
1% Sodium pyruvate solution
1% Nonessential amino acids solution
1% Penicillin/streptomycin

2 M CaCl$_2$
CaCl$_2$ (dihydrate, cell culture grade, Sigma), 2 M.
Dissolve in autoclaved H$_2$O.
Filter-sterilize through 0.2-μm filter.
Prepare fresh each time.

Puck's saline
NaCl .... 140 mM
KCl .... 50 mM
KH$_2$PO$_4$ .... 1 mM
Na$_2$HPO$_4$ .... 1 mM
Glucose .... 6 mM

Adjust to pH 7.4.
Filter-sterilize.
Store at 4°C.

HEPES-buffered saline (HBS), 2×
NaCl .... 0.3 M
Na$_2$HPO$_4$ .... 2.8 mM
HEPES .... 50 mM

Dissolve in autoclaved H$_2$O.
Adjust pH to 7.12.
Filter-sterilize through 0.2-μm filter.
May be stored at −20°C for up to 1 month.

15% glycerol in HBS

Glycerol, (molecular biology grade) .... 15% v/v
H$_2$O .... 35% v/v
HBS 2× .... 50% v/v

Filter-sterilize through 0.2-μm filter.
May be stored at 4°C for up to 1 month.

4.2. Nonsterile Reagents

The following solutions are required for tissue harvesting and for gene analysis by CAT and β-gal assays. Also listed are several reagents for tissue immunohistochemistry requiring advanced preparation.

Phosphate-buffered saline (PBSA), pH 7.4

NaCl .... 140 mM
KCl .... 2.7 mM
Na$_2$HPO$_4$ .... 10 mM
KH$_2$PO$_4$ .... 1.8 mM

Phosphate-buffered saline (PBSA), pH 7.4, 10×

NaCl .... 1.40 M
KCl .... 27 mM
Na$_2$HPO$_4$ ...... 0.1 M
KH$_2$PO$_4$ .... 18 mM

TENs buffer, 1×

NaCl, ... 0.1 M
Tris HCl .... 10 mM
EDTA, pH 8.0 ...... 1.0 mM

0.25 M Tris HCl, pH 7.8

Tris .... 0.25 M

Adjust pH to 7.8 with HCl.

Mg buffer, 100×

MgCl$_2$ .... 1 M
β-Mercaptoethanol .... 5 M

Na$_2$HPO$_4$/NaH$_2$PO$_4$ solution, 0.1 M, pH 7.4

Na$_2$HPO$_4$, 84 mM .... 11.3 g
NaH$_2$PO$_4$, 17.5 mM ...... 2.07 g

Bring volume up to 1 liter with H$_2$O.

β-galactosidase enzyme dilution

1000 U β-galactosidase (Boehringer Mannheim)

Bring volume up to 333 μl with 0.1 M Na_2HPO_4/NaH_2PO_4 solution, pH 7.4.

ONPG (13 mM 2-nitrophenyl-β-D-galactopyranoside), 1×
4 g of ONPG

Bring volume up to 25 ml with 0.1 M Na_2HPO_4/NaH_2PO_4 solution, pH 7.4.

Acetyl coenzyme A solution, 4 mM
1.762 mg of acetyl coenzyme A, lithium salt

Bring volume up to 500 ml with H_2O.
Store at 4°C.

Paraformaldehyde, 4% in PBSA
4 g paraformaldehyde (EM grade)

Bring volume up to 50 ml with H_2O.
Stir and heat to 65°C (solution appears cloudy).
Add 10 μl 10 N NaOH to clear.
Filter through #1 Whatman paper.
Add 10 ml of 10× PBSA.
Adjust the volume to 100 ml with H_2O.
Store at 4°C for up to 24 h.

Paraffin tissue-embedding media (Paraplast X-TRA, EMS)
Melt sufficient paraffin quantities at 55–60°C 1–2 days prior to use.
Filter through #1 Whatman paper to remove debris.
Store in oven at 55–60°C until ready to use.

4.3. Sterile Supplies

Pasteur pipettes, 9″
Tissue culture petri dishes, 100 mm; 35 mm 6-well
Conical tubes, 50 ml
Eppendorf tubes

4.4. Nonsterile Supplies

Silane-treated slides
Wash slides 7× by soaking in dish detergent.
Rinse in running H_2O for 1 h.
Rinse in acetone.
Immerse in 2% 3-aminopropyl-triethoxy-silane in acetone.
Rinse in running H_2O for 5 min.
Dry overnight at room temperature.

Microtome
Glass slides, micro slides, 0.96–1.06 mm thickness
Coverslips, cover glass #1
Slide heating tray
Bibulous paper
Staining racks and dishes
Humidified chamber

5. TRANSIENT TISSUE TRANSFECTION PROTOCOLS

5.1. Introduction

A transient transfection procedure of whole embryonic chick lungs was developed to determine the activity of various elastin promoter deletion constructs in a whole organ model system [James, 1997; James et al; 1998]. Preliminary trials employing calcium phosphate coprecipitation, liposome (Lipofectin, Gibco BRL) and electroporation methods of gene transfer were conducted to determine the optimum transient transfection conditions. These analyses employed numerous modifications of each procedure in order to obtain the optimal level of transfection efficiency while maintaining the proper integrity of the tissue. A protocol using calcium phosphate coprecipitation transfection [Chen and Okayama, 1987] was ultimately selected because it resulted in quantitatively reproducible CAT assays while maintaining histologically intact lung tissue structure.

In our hands, the liposome-mediated method of gene transfer [Felgner et al., 1987] was not shown to be as effective as the calcium phosphate method for introducing the elastin promoter/reporter construct into tissue. It should be noted that although immunohistological staining of the reporter CAT protein in liposome-transfected tissue produced reporter distribution patterns of both the endogenous elastin and CAT proteins qualitatively similar to those in calcium phosphate-transfected tissue (Figs. 9.2 and 9.3), the magnitude of transfection efficiency was less. This was determined by comparing the activity levels of CMV-β-galactosidase per microgram of protein tissue extract for each transfection procedure. Transfection efficiency was demonstrated to be 50–180% greater using the calcium phosphate coprecipitation method as compared to the liposome-mediated method under the conditions tested. Our attempts at tissue transfection by electroporation was not successful, since this technique resulted in moderate to severe structural damage to the lung tissue.

The *in vitro* transfection method described in the following protocol was optimized for maximum gene expression of the p2.2CAT human elastin promoter/CAT reporter construct [Wolfe et al., 1993] in day 18 embryonic chick lung organ culture. In parallel, pCAT-Basic (Promega), which lacks eukaryotic promoter and enhancer sequences, was transfected for use as a negative control. Transfection efficiencies were assessed by determining activity levels of the cotransfected pCMV-β-gal reporter vector, which expresses the full-length *E. coli* β-galactosidase gene from the constitutively active human cytomegalovirus immediate early gene promoter (CLONTECH Laboratories, Inc., Palo Alto, CA).

5.2. Calcium Phosphate Coprecipitation Tissue Transfection

5.2.1. Tissue culture preparation

Protocol 5.2.1

Reagents and Materials
- Embryonated eggs, 18 days
- DMEM/10FB
- Forceps
- Scalpels
- Petri dishes, 100 mm
- Multi-well plate, 6 × 35 mm

Protocol
(a) Excise the tissue to be transfected (e.g., day 18 embryonic chick lungs) using standard sterile technique [Freshney, 1994].
(b) Place the tissue samples in a 100-mm petri dish and wash residual blood from tissue with DMEM/10FB.
(c) Transfer two chick lungs into the 35-mm well of a 6-well plate, each well containing 4.5 ml of fresh DMEM/10FB; one well for each experimental condition. (*Wet weight of each embryonic day 18 chick lung is approximately 80 mg.*)
(d) Incubate tissue in a humidified 5% CO_2 air incubator at 37°C for 2 h.

5.2.2. DNA-precipitate preparation

Protocol 5.2.2

Reagents and Materials
Aseptically prepared
- CAT reporter construct
- pCMV-β-gal expression vector
- Eppendorf tube
- Lung cultures in 35-mm, 6-well plate
- UPW

- CaCl$_2$, 2 M
- HBS, 2×

Protocol
(a) Combine 65 μg of appropriate CAT reporter construct and 20 μg pCMV-β-gal expression vector, for each 35-mm well of the plate, in sterile Eppendorf tubes. (*Different DNA concentrations should be tested for optimal gene expression.*)
(b) Add sterile UPW to the above DNA to a final volume of 657 μl and vortex. (*Add no DNA for control mock transfected samples.*)
(c) Add 93 μl of 2 M CaCl$_2$ to the DNA solution, to make a volume of 750 μl, and mix.
(d) Add the DNA/CaCl$_2$ mixture dropwise, slowly, into a 50-ml conical tube containing 750 μl of 2× HBS with air bubbling through to minimize precipitation in large clumps.
(e) Allow the DNA/Ca$_3$(PO$_4$)$_2$ precipitate to stand at room temperature for 10 min.

5.2.3. Transfection procedure

Protocol 5.2.3

Reagents and Materials
- DNA/Ca$_3$(PO$_4$)$_2$ solution from Protocol 5.2.2
- Pasteur pipettes
- Puck's saline
- Glycerol, 15% in HBS
- PBSA

Protocol
(a) Add DNA/Ca$_3$(PO$_4$)$_2$ solution (1.5 ml total volume) dropwise to each 35-mm well, containing 2 chick lung samples, to give a final volume of 6.0 ml.
Note: Variations on this step include incubating DNA with tissue under agitating conditions using a platform shaker.
(b) Incubate for 16–24 h at 37°C in humidified 5% CO$_2$ incubator.
(c) Remove medium from tissue using a sterile Pasteur pipette.
(d) Wash tissue twice with 3–5 ml of Puck's saline.
(e) Add 1 ml of 15% glycerol in HBS for 30 s to shock the tissue.
(f) Remove the 15% glycerol in HBS solution from the culture dish and rinse the tissue twice with Puck's saline.
(g) Add 6 ml of fresh DMEM/10FB to chick lungs.
(h) Incubate for 24 h.
Note: Vary incubation time for maximal gene expression.
(i) Remove the medium from the culture dishes.
(j) Rinse lung tissue twice with 5 ml of PBSA.

(k)　Transfer one lung from each sample condition to a 50-ml conical tube containing 4% paraformaldehyde in PBSA for analysis of reporter activity by immunohistochemical methods.

(l)　Transfer the second transfected lung sample to an Eppendorf tube containing 0.5 ml 1× TENs buffer for tissue lysate preparation.

5.2.4.　Harvesting tissue lysate

Protocol 5.2.4

Reagents and Materials
Nonsterile
- TENs buffer, 1×
- Vortex mixer
- Microfuge
- Sonicator
- Tris HCl, 0.25 M, pH 7.8
- Ice

Protocol
(a)　Disrupt the tissue sample in 0.5 ml 1× TENs buffer by vigorous vortexing and pipetting.

(b)　Microcentrifuge the tissue suspension at 12,000 g for 3 min.

(c)　Resuspend the tissue pellet in 150 μl cold 0.25 M Tris HCl, pH 7.8.

(d)　Lyse cells by sonicating three times for 10-s intervals, returning the sample to ice each time.

(e)　Microcentrifuge tissue debris at 12,000 g for another 3 min at 4°C.

(f)　Analyze the extract for CAT and β-gal activity and for total protein content.

(g)　Store the tissue lysates at −20°C.

6.　GENE REPORTER ASSAYS

Control and transfected tissue may be processed by several methods to test for reporter gene activity. Parallel tissue samples for each experimental condition may either be subjected to tissue fixation for use in immunohistological analysis or harvested for β-galactosidase and CAT activity, essentially as described by Herbomel et al. [1984]. Total protein content should also be determined. Conventional thin-layer chromatography (TLC) CAT assays are conducted to measure reporter activity levels using tissue extracts obtained from transfected lung samples. In addition, immunohistochemical staining methods are applied to transfected tissue sections. CAT antibodies are used to detect the cellular

distribution pattern and specificity dictated by the elastin promoter sequences.

6.1. β-Galactosidase Assays

Protocol 6.1

Reagents and Materials
Nonsterile
- Mg buffer, 100× (see Section 4.2)
- ONPG, 1× (see Section 4.2)
- Na_2HPO_4/NaH_2PO_4 solution, 0.1 M, pH 7.4
- Na_2CO_4, 1M (see Section 4.2)
- Bacterial β-galactosidase

Protocol
(a) Mix 30 μl of tissue extracts in duplicate in Eppendorf tubes with
the following reagents:
  (i) 3 μl 100× Mg buffer
  (ii) 66 μl 1× ONPG
  (iii) 201 μl 0.1 M Na_2HPO_4/NaH_2PO_4 solution, pH 7.4, for a
  total volume of 300 μl
(b) Delete tissue extract from negative control assay.
(c) Delete tissue extract and add 1 μl (3 U) of bacterial β-
galactosidase enzyme dilution for positive control assay.
(d) Incubate assays at 37°C for 1–2 h.
(e) Terminate reactions with 0.5 ml 1 M Na_2CO_3.
(f) Assess activity by spectrophotometric analysis at 410 nm.

6.2. CAT (Chloramphenicol Acetyltransferase) Assays

Protocol 6.2

Reagents and Materials
Nonsterile
- BCA protein assay kit or equivalent
- Acetyl coenzyme A solution, 4 mM
- CAT enzyme
- [^{14}C]Chloramphenicol
- Tris HCl, 0.25 M, pH 7.8
- Ethyl acetate (HPLC grade)
- Silica gel thin-layer chromatography (TLC) plates
- Chloroform/methanol (95:5).

Protocol
(a) Determine the total protein concentrations of the tissue extracts
by BCA protein assay or similar method according to the manufacturer's instructions.

(b)   Mix a constant amount of protein (100–300 μg) from each tissue lysate sample with the following components:

    20 μl 4 mM acetyl coenzyme A solution
    2 μl (0.05 μCi) [^{14}C]-chloramphenicol
    0.25 M Tris HCl, pH 7.8, for a total volume of 180 μl

(c)   Delete tissue lysate from the negative CAT control assay.

(d)   Delete tissue lysate and add 2700 U of CAT enzyme for the positive control assay.

(e)   Incubate the assays at 37°C for 8 h.

(f)   Recover acetylated [^{14}C]-chloramphenicol derivatives from each sample by performing three ethyl acetate extractions.

(g)   Evaporate the solvent from the samples in a fume hood overnight.

(h)   Reconstitute the samples in 15 μl of ethyl acetate.

(i)   Spot samples onto silica gel plates.

(j)   Separate the [^{14}C]-acetylated chloramphenicol derivatives by TLC in 200 ml of chloroform/methanol (95:5).

Note: Figure 9.1 shows a representative CAT assay using tissue extract from transiently transfected p2.2CAT/CMV-β-gal chick lungs.

(k)   Determine the CAT reporter activity by quantitating percent CAT conversion using densitometric analysis of autoradiographic plates and/or phosphor-imager analysis of phosphor-imager TLC-exposed plates.

(l)   Normalize the CAT value in each sample (x) to β-galactosidase activity level (y) to compensate for transfection efficiency as follows: the highest value of y among all samples will be considered as 1 and the rest as fraction of 1. For each sample the value x will be divided by its corresponding fraction value or by 1 (for the sample with highest β-galactosidase activity).

6.3.   Immunohistochemistry

6.3.1.   Tissue fixation and slide preparation

Protocol 6.3.1
Fix transiently transfected chick lung tissue in paraformaldehyde and embed in paraffin following the protocol of Sassoon et al. [1988].

Reagents and Materials
Nonsterile materials
• PBSA, 1×
• Paraformaldehyde, 4% in PBSA
• NaCl, 0.85%
• 50% ethanol in 0.85% NaCl
• 70% ethanol in water

James et al.

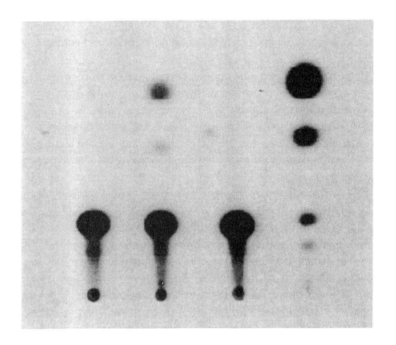

Fig. 9.1. CAT assay of calcium phosphate transfected chick lung tissue. Equivalent protein concentrations from p2.2CAT and mock transfected (no DNA) tissue lysates were assayed by standard TLC methods. Autoradiograph shows resulting acetylated chloramphenicol conversion products, (+) and p2.2CAT.

(-)   p2.2CAT   mock   (+)

- 85% ethanol in water
- 95% ethanol in water
- 100% ethanol
- Xylene (use in fume hood)
- Xylene:paraffin wax, 1 : 1
- Paraffin wax
- Ethanol in H_2O, 0.2%
- Silane-treated slides
- Bibulous paper

Nonsterile equipment

- Microtome
- Vacuum pump

Protocol

(a) Remove the media from cultured tissue and wash the tissue twice in 1× PBSA.

(b) Add the tissue to 4% paraformaldehyde in PBSA at approximately 0.5 cm³ of tissue per 50 ml of fixative solution.

(c) Continue fixing tissue for 20 h at 4°C *in vacuo*.

(d) Remove the fixative from tissue by washing in PBSA and replacing with 0.85% NaCl at 4°C for 30 min.

   Note: Perform these and subsequent washing steps (b through f) in 50-ml conical tubes with occasional agitation using wash volumes of approximately 30 ml.

(e) Transfer the tissue into the following ascending ethanol series to dehydrate the tissue by gradually replacing the saline with alcohol:

   (i) 1:1 saline:100% ethanol wash at room temperature for 20 min

   (ii) 70% ethanol wash, twice at room temperature for 30 min each

   (iii) 85% ethanol wash at room temperature for 45 min

   (iv) 95% ethanol wash at room temperature for 45 min

   (v) 100% ethanol wash, twice at room temperature for 45 min each

   (vi) 100% ethanol wash overnight at 4°C *in vacuo*

(f) Transfer the tissue into the following washes to replace the ethanol with xylene and subsequently the xylene with paraffin tissue-embedding media.

   (i) 100% xylene wash, twice at room temperature for 30 min each

   (ii) 1:1 xylene:paraffin wax solution for 45 min at 60°C

   (iii) 100% paraffin, three times for 20 min each at 60°C

(g) Orient the tissue in a plastic mold during the third 100% paraffin wash.

(h) Keep the tissue-paraffin mold at room temperature overnight to allow the paraffin to solidify.

(i) Remove the embedded-tissue block mold.

(j) Mount the tissue-paraffin block on a microtome.

(k) Slice the tissue in sections 4–6 μm thick.

(l) Warm silane-treated slides to 48°C on a slide heating tray.

(m) Place 500 μl of 0.2% ethanol in H_2O solution onto glass slides.

(n) Place paraffin-embedded tissue sections on the pre-warmed ethanol/H_2O solution.

(o) Let the tissue sections stand approximately 3–5 min in the ethanol/H_2O solution.

   Note: This process allows paraffin-embedded tissue sections to flatten and unfold to their original configuration, which may have been distorted during the slicing process.

(q) Pour excess ethanol/H_2O solution from the slide while retaining the tissue section.

(r) Place water-moistened bibulous paper over the tissue section and press lightly with a rubbing motion so that the tissue adheres to the slide.

(s)   Place the slides upright at room temperature overnight to secure tissue sections to slides.

(t)   Store the tissue sections at 4°C.

6.3.2.   Tissue preparation and antibody staining

Protocol 6.3.2

Transfected tissue sections are prepared for indirect immunofluorescent staining as described by Gruskin-Lerner and Trinkaus-Randall [1991]. The paraffin-embedded tissue-sectioned slides are pretreated to remove the paraffin and permeabilized to allow for the cellular integration of antibody probes.

Reagents and Materials
Nonsterile
- Xylene
- Ethanol, 100%
- Ethanol, 95%
- Ethanol, 70%
- UPW
- PBSA
- BSA, 3% in PBSA
- Secondary immunofluorescent-conjugated antibody
- Slides and coverslips
- Slide rack
- Staining dishes
- Humidifier chamber
- Anti-fade solution, e.g., Vectastain
- Fingernail polish

Protocol
(a)   Immerse the slide rack containing tissue-sectioned slides into staining dishes containing the following solutions:
    (i)   100% xylene, twice for 5 min each
    (ii)   100% ethanol three times for 5 min each
    (iii)   95% ethanol twice for 5 min each
    (iv)   70% ethanol for 5 min
(b)   Rinse the slides gently under running UPW for 1 min.
    Note: Do *not* allow the tissue sections to air dry at any point during this procedure.
(c)   Transfer the tissue sections to a staining dish containing PBSA, twice for 5 min each.
(d)   Encircle each tissue section with a wax pencil to prevent the contamination and leakage of different antibodies.
(e)   Position the tissue-sectioned slides horizontally.

(f)   Add 50 ml of 3% BSA/PBS solution to tissue sections for 30 min at room temperature to block nonspecific binding.

(g)   Remove the blocking solution.

(h)   Add 50 ml of the appropriate primary antibodies to tissue sections. Use antibodies diluted in 1% BSA/PBS as follows:

    STI elastin primary antibody   1 : 20
    anti-CAT-digoxigenin   1 : 10
    mouse anti-β-galactosidase   1 : 50
    (See Fig. 9.2)

(i)   Transfer the horizontally positioned slides to a humidified chamber to prevent evaporation.

(j)   Let the slides stand for 24 h at 4°C.

(k)   Rinse the slides of primary antibody with running PBSA for 1 min.

(l)   Submerge the slides in PBSA twice, for 30 min each, at room temperature.

(m)   Transfer the slides into staining dishes containing 3% BSA/PBSA.

(n)   Let the slides stand for 24 h at 4°C.

(o)   Transfer the slides to a horizontal position in a humidified chamber.

(p)   Add 50 ml of the appropriate secondary immunofluorescent-conjugated antibody, diluted in 1% BSA/PBS as follows:

    sheep anti DIG-fluorescein isothiocynate (FITC)   1 : 5
      conjugated Fab fragments
    FITC-conjugated goat anti-mouse IgG poly-   1 : 50
      clonal antibody
    (See Fig. 9.2)

(q)   Incubate the tissue sections at 37°C for 1 h in the humidified chamber.

(r)   Wash unbound secondary antibodies from the tissue by applying running PBSA for 1 min.

(s)   Add a drop of anti-fade solution to tissue sections to prolong the immunofluorescent signal.

(t)   Cover the tissue sections with glass coverslips and attach the coverslips to the slides using fingernail polish.

(u)   Visually examine the tissue sections using confocal scanning laser microscopy (CSLM) or similar immunofluorescent microscopy.

Negative controls for immunohistochemistry include those in which either the primary or secondary antibody is omitted on serial tissue sections. Negative controls for β-gal and CAT staining also include mock and non-transfected tissue sections exposed to both the primary and secondary antibodies specific for each antigen being tested. Tissue sections that are analyzed by CSLM are compared with regard to staining intensities for a given antibody at similar parameters for laser

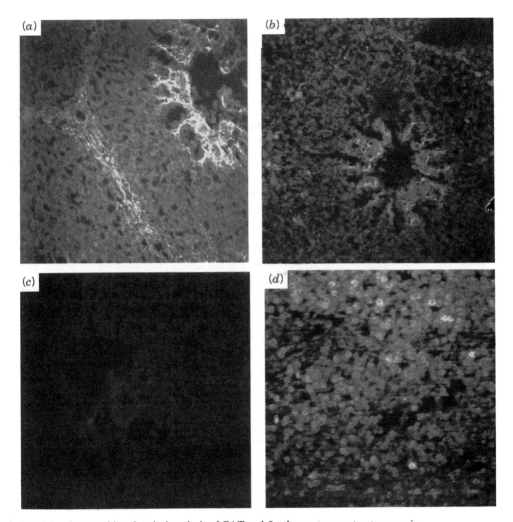

Fig. 9.2. Immunohistochemical analysis of CAT and β-gal reporter construct expression in calcium phosphate transfected chick lung tissue. Antibodies were diluted in 1% BSA/ PBS. Panel *a* is a mock transfected tissue sample which demonstrates the endogenous elastin protein expression pattern in specific regions of the chick lung tertiary bronchial unit. The ST1 elastin primary antibody (Foster et al., 1976) was diluted 1:20 as above and was detected with affinity purified goat anti-rabbit IgG antibody conjugated to rhodamine (Boehringer Mannheim). Panels *b* and *c* are lung tissue sections transfected with reporter constructs p2.2CAT/CMV-β-gal and pCAT-Basic/CMV-β-gal, respectively. Each tissue section was treated with anti-CAT-digoxigenin (DIG; Boehringer Mannheim), a polyclonal antibody from sheep that is specific for CAT type I (used at a 1:10 dilution). This primary anti-CAT antibody was used sequentially with secondary sheep anti-DIG-fluorescein iso-thiocyanate (FITC) conjugated Fab fragments (Boehringer Mannheim) at a 1:5 dilution. Panel *d* shows a p2.2CAT/CMV-β-gal transfected tissue section that was treated with affinity purified mouse anti-β-galactosidase antibody (Sigma), diluted 1:50. This was detected by a FITC-conjugated goat anti-mouse IgG polyclonal antibody (Boehringer Mannheim), diluted 1:50. Images were analyzed using CLSM. Scale is 0.25″ in picture = 10 microns.

power, voltage, and aperature. Representative areas and controls are scanned for each experiment.

7. RESULTS

Immunohistochemical staining of the CAT enzyme in calcium phosphate transiently transfected lung tissue sections is used to determine the expression patterns, activity level, and specificity of the human elastin promoter p2.2CAT reporter construct. These results are compared to the distribution patterns of the endogenous elastin protein, the pCAT-Basic reporter construct, and the constitutively active CMV-β-galactosidase protein product (Fig. 9.2). Our results show that the p2.2CAT expression construct colocalizes with the endogenous elastin protein, while the transfected β-galactosidase enzyme is present in a diffuse pattern throughout the lung tissue, reflecting the strong and general activity of the CMV promoter. The pCAT-Basic expression construct does not express the CAT protein.

Both methods of transient tissue transfection—calcium phosphate co-precipitation and the liposome technique—demonstrate similar p2.2CAT expression patterns (Fig. 9.3a and b). Mock transfected tissue sections shown in Panels d and e demonstrate the specificity of CAT labelling using both the calcium phosphate co-precipitation and liposome methods.

ACKNOWLEDGMENTS

This work was supported by National Institute of Health Grant HL-46902. MF James was supported by Training Grant HL-07429.

Fig. 9.3. Comparison of calcium phosphate coprecipitation and liposome-mediated methods of gene transfer. Panels a and b show p2.2CAT distribution patterns from transfected lung tissue using the calcium phosphate coprecipitation and lipofection methods, respectively. Panels d and e show tissue sections that were mock transfected (negative control sections) using the calcium phosphate coprecipitation and lipofection methods, respectively. Tissue sections shown in Panels a, b, d, and e were all treated with the primary antibody, anti-CAT-DIG, and the secondary antibody, anti-DIG-FITC, as described in Fig. 9.2. Panel c shows a serial lung tissue section from the same tissue as shown in Panel A (p2.2CAT transfected). This section was treated as a negative control using 1% BSA/PBS as the primary antibody and anti-DIG-FITC as the immunofluorescent secondary antibody. Images were analyzed using CLSM. Scale is 0.25″ in picture = 10 microns.

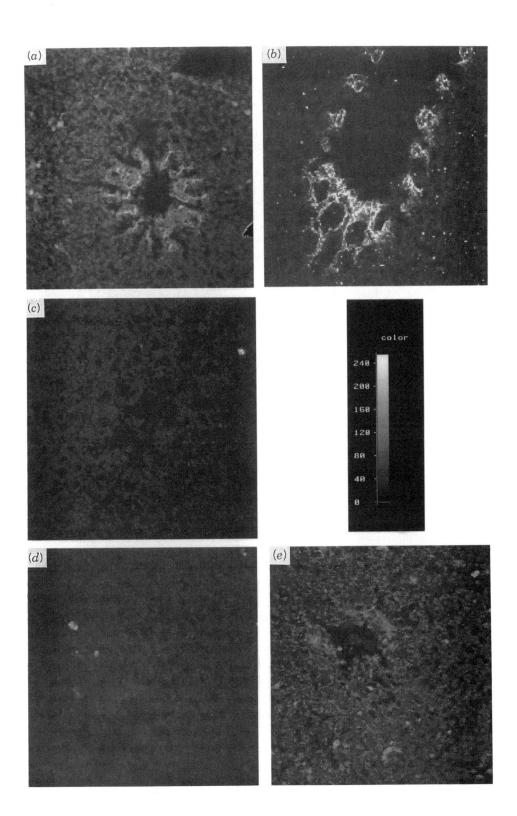

REFERENCES

Arnold D, Feng L, Kim J, Heintz N (1994): A strategy for the analysis of gene expression during neural development. Proc Natl Acad Sci USA 91: 9970–9974.

Ben-Drohr I, Havazelet N, Vardimon L (1993): Proc Natl Acad Sci USA 90: 1117–1121.

Berko-Flint Y, Levkowitz G, Vardimon L (1994): Involvement of c-Jun in the control of glucocorticoid receptor transcriptional activity during development of chicken retinal tissue. EMBO J 13: 646–654.

Chen C, Okayama H (1987): High-efficiency transformation of mammalian cells by plasmid DNA. Mol Cell Biol 7: 2745–2752.

Felgner PL, Gadek TR, Holm M, Roman R, Chan HW, Wenz M, Northrup JP, Ringold GM, Danielson M (1987): Lipofection: a highly efficient, lipid-mediated DNA transfection procedure. Proc Natl Acad Sci USA 84: 7413–7417.

Fisher SA, Walsh K, Forehand CJ (1996): Characterization of cardiac gene cis-regulatory elements in the early stages of chicken heart morphogenesis. J Mol Cell Cardiol 28: 113–122.

Foster JA, Knaack D, Faris B, Moscaritolo R, Salcedo L, Skinner M, Franzblau C (1976): Development of a specific immunological assay for tropoelastin and its application to tissue culture studes. Biochem Biophys Acta 446: 51–55.

Freshney RI (1994): "Culture of Animal Cells—A Manual of Basic Technique," New York: Wiley-Liss, pp 129–133.

Gruskin-Lerner L, Trinkaus-Randall V (1991): Localization of integrin and syndecan *in vivo* in a corneal epithelial abrasion and keratectomy. Curr Eye Res 10: 75–85.

Herbomel P, Bourachot B, Yaniv M (1984): Two distinct enhancers with different cell specificities coexist in the regulatory region of polyoma. Cell 39: 653–658.

James MF (1997): "The Regulation of Elastogenesis During Chick Lung Development." PhD dissertation, Boston University School of Medicine.

James MF, Rich CB, Ginsburg CD, Foster JA (1998): *In vivo* analysis of elastin gene expression in the developing chick lung (manuscript in preparation).

Kuhn C (1978): Ultrastructure and cellular function in the distal lung. In Thurlbeck W (ed): "The Lung: Structure, Function and Disease." Baltimore: Williams & Wilkins, pp 3–48.

Linser P, Moscona AA (1979): Induction of glutatmine synthetase in embryonic neural retina: localization in Muller fibers and dependence on cell interactions. Proc Natl Acad Sci USA 76: 6476–6480.

Linser P, Perkins MS (1987): Regulatory aspects of the in vitro development of retinal Muller glical cells. Cell Growth Differ 20: 189–196.

Moscona AA, Linser P (1983): Developmental and experimental changes in retinal glial cells: cell interactions and control of phenotype expression and stability. Curr Top Dev Biol 18: 155–188.

Pu H, Young AP (1990): Glucocorticoid-inducible expression of a glutamine synthetase-CAT-encoding fusion plasmid after transfection of intact chicken retinal explant cultures. Gene 89: 259–263.

Sassoon DA, Garner I, Buckingham M (1988): Transcripts of alpha-cardiac and alpha-skeletal actins are early markers for myogenesis in the mouse embryo. Development 104: 155–164.

Takeshita S, Gal D, Leclerc G, Geoffrey Pickering J, Riessen R, Weir L, Isner JM (1994): Increased gene expression after liposome-mediated arterial gene transfer associated with intimal smooth muscle cell proliferation. J Clin Invest 93: 652–661.

Wolfe LB, Rich CB, Danana Goud H, Terpstra AJ, Bashir M, Rosenbloom J, Sonenshein GE, Foster JA (1993): Insulin-like growth factor-1 regulates transcription of the elastin gene. J Biol Chem 268: 12418–12426.

Zhang H, Young AP (1991): A single upstream glucocorticoid response element juxtaposed to an AP1/ATF/CRE-like site renders the chicken glutamine synthetase gene hormonally inducible in transfected retina. J Biol Chem 266: 24332–24338.

APPENDIX: MATERIALS AND SUPPLIERS

Materials	Suppliers
Acetyl coenzyme A, lithium salt	Pharmacia
3-Aminopropyl-triethoxy-silane	Sigma
Anti-Fade Solution, SlowFade Anti Fade Kit,	Molecular Probes
Bibulous paper	Fisher
BCA protein assay	Pierce Chemical
β-Galactosidase	Boehringer Mannheim
Bovine serum albumin, BSA, (fraction V),	Sigma
CaCl$_2$ (dihydrate, cell culture grade)	Sigma
CAT enzyme	Pharmacia
^{14}C-Chloramphenicol, (D-*threo*-[dichloracetyl-1-^{14}C] chloramphenicol) 250 mCi/ml	Amersham
Coverslips, glass #1	Corning
Dulbecco's modified Eagle's medium (DMEM)	JRH Biosciences
Fetal bovine serum (FBS)	Sigma
Glycerol (molecular biology grade)	IBI
Human cytomegalovirus immediate early gene promoter	CLONTECH Laboratories
Liposome reagent, LIPOFECTIN reagent	Life Technologies
Microscope slides, 0.96–1.06 mm	Corning
Non-essential amino acids solution	Gibco BRL
Paraformaldehyde (EM grade)	Polysciences
pCAT-basic	Promega
Penicillin/streptomycin	Gibco BRL
Petri dishes and other TC plastics	Corning-Costar
Silica gel TLC plates	JT Baker
Sodium pyruvate solution	Gibco BRL
Staining racks and dishes	Fisher
Wax pencil, Pap pen	EMS-Electron Microscopy Sciences

10

DEAE-Dextran Transfection

John J. Schwartz and Robert Rosenberg

Massachusetts Institute of Technology, Department of Biology, 77 Massachussetts Avenue, Building 68-480, Cambridge, Massachusetts 02139-4307

I. INTRODUCTION AND BACKGROUND

The DEAE-dextran technique was originally used to facilitate the transfer of viral genomic RNAs into host cells [McCutchen and Pagano, 1968]. Today, DEAE-dextran transfection is typically used for the transient expression of plasmid-borne sequences in cultured cell-lines [Sussman et al., 1984]. This allows for the expres-

DNA Transfer to Cultured Cells, Edited by Katya Ravid and R. Ian Freshney.
ISBN 0-471-16572-7 © 1998 Wiley-Liss, Inc.

sion cloning of cDNAs, analysis of enhancer and promoter function, identification of transcriptional and translational control elements, analysis of mRNA stability and protein function [Fujita et al., 1986; Reeves et al., 1985; Seldon et al., 1986; Pearson et al., 1995; Orellana et al., 1994]. The DEAE-dextran method is not as useful for the establishment of stable transfectant cell lines as are other techniques like calcium phosphate precipitation, lipofection and electroporation [Holter et al., 1989; Maniatis et al., 1989; Schwachtgen et al., 1994; Kahn et al., 1992; Puchalski et al., 1992]. However, it can offer several advantages over other methods of transfection:

(i)   A small amount of plasmid DNA is required compared to other transfection methods (typically 0.1–5.0 μg of DNA).

(ii)   Many cultured cell types (e.g., COS, CV-1, BSC-1, Lta, Raji, Jurkat-111, etc.) can be transfected by this method [Fujita et al., 1986; Puchalski et al., 1992; Ausubel et al., 1995a].

(iii)   Both adherent and suspension cultures can be transfected [Ausubel et al., 1995a].

(iv)   Once optimized, DEAE-dextran transfection can be simply scaled up to transfect a larger number of cells.

With COS cells, DNA transfection mediated by DEAE-dextran is a reliable, simple and efficient technique. Some cell types (e.g., primary cell isolates in culture) are often difficult to transfect efficiently [Felgner and Rhodes, 1991]. In these cases, DNA transfer by other methods (e.g., electroporation, lipofection, calcium phosphate, or virus-mediated gene transfer) should be employed. With the COS-7 (African green monkey kidney) cell line, there are several factors that greatly influence the success of DEAE-dextran transfection:

(i)   Concentration of the DEAE-dextran.

(ii)   Length of exposure to the DEAE-dextran transfection cocktail (see below).

(iii)   Time and concentration of chloroquine to which the cells are exposed.

(iv)   Purity of the DMSO.

(v)   Gentle washing after the DMSO shock.

(vi)   The growth state of the COS-7 cells prior to transfection [Corsaro and Pearson, 1981; Luthman and Magnusson, 1983; Sussman and Milman, 1984; Ausubel et al., 1995a; Schwartz personal observations].

There are many different protocols that have been reported using DEAE-dextran. Each cell line, or cloned cell strain, is affected differently by this procedure; therefore, pilot titration experiments need to be undertaken to determine the best conditions for the cells being transfected. The following protocols have proven successful for the authors and should be used as a starting point for further improvements. Included with the protocols is a list of reagents and suppliers used by the authors; however, acceptable results can surely be obtained with reagents and disposables from a variety of manufacturers.

2. PREPARATION OF REAGENTS AND MEDIA

Solution 1: DEAE-Dextran in DMEM
Aseptically prepared
Dulbecco's modified Eagle's medium
(without serum) .... 100 ml
DEAE-dextran .... 25 mg
1 M Tris-base, pH 7.4 .... 5 ml
Total volume .... 105 ml

Mix to dissolve with a stir bar, at 37°C for 15–30 min. Sterilize the solution through a 0.22-μm filter. Make the solution up fresh each time. (This volume is sufficient for 50–100 transfections of 35-mm tissue culture dishes.)

Solution 2: 100× Chloroquine-diphosphate Stock Solution
Aseptically prepared
Chloroquine-diphosphate (MW 515.5) ...... 25 mg
Distilled deionized water 5ml

Sterilize the solution through a 0.22-μm filter and store at −20°C wrapped in aluminum foil. This solution can be kept for up to 6 months if frozen and protected from exposure to light. (This volume is enough for 250–500 transfections of 35-mm tissue culture dishes.)

Solution 3
Aseptically prepared
1× PBSA .... 54 ml
DMSO .... 6 ml
Total volume .... 60 ml

Make this solution up fresh each time and allow it to cool for at least 15 min before use. Either use sterile 1× PBSA, or sterilize the solution through a 0.22-μm filter before use. (This volume is enough for 60 transfections in 35-mm dishes.)

Transfection Cocktail

To 105 parts of Solution 1, add 1 part of Solution 2. It is important that the transfection cocktail solution be mixed just prior to use and be protected from light. Make up 1–2 ml of transfection cocktail for each 35-mm dish to be transfected. Make up only as much as is needed and do not use this solution after 24 h.

Suspension TBS (STBS)

Tris-HCl, pH 7.4 .... 25 mM
NaCl .... 137 mM
KCl .... 5 mM
Na_2HPO_4 .... 0.6 mM
$CaCl_2$ .... 0.7 mM
$MgCl_2$ .... 0.8 mM

10 mg/ml DEAE-Dextran in STBS

Prepare a 10 mg/ml stock solution in $1\times$ STBS and filter-sterilize through a 0.2-μm filter. Store the stock solution in a sterile tube at 4°C. Do not freeze this solution, as the DEAE-dextran may be degraded by ice crystal formation.

Make up this solution in distilled deionized water and filter-sterilize through a 0.2-μm filter. A $10\times$ stock of this solution can be prepared and diluted to the required strength with distilled deionized sterile water prior to each experiment.

Fixation Solution for β-Galactosidase

Glutaraldehyde (2.5 M stock) ....1% vol/vol
$NaHPO_4$, pH 7.4 .... 0.1 M
PBSA
$MgCl_2$ .... 1 mM

X-Gal Stock Solution

X-Gal, 40 mg/ml in N,N-dimethylformamide (DMF)
(Use glass or polypropylene tubes to contain neat DMF solutions.)

X-Gal Staining Solution for β-Galactosidase

Potassium ferrocyanide, 55 mM .... 0.5 ml
Potassium ferricyanide, 55 mM .... 0.5 ml
$MgCl_2$, 1 M .... 10 μl
PBSA, pH 7.4 .... 8.74 ml
X-Gal .... 0.25 ml

3. TRANSFECTION PROCEDURES

3.1. Adherent Cells

An overview of the time-line for the procedure is as follows:

Day 0   Plate 100,000 cells per 35-mm dish.
Day 1   Replace the medium with DMEM, 10% Nu-Serum.
Day 2   Transfect cells.
Day 3–5   Harvest cells or supernatant, and analyze samples.

Protocol 3.1

Reagents and Materials

Aseptically prepared

- Plasmid: Prior to beginning this protocol, at least 100 μg of plasmid DNA should be prepared via the QIAGEN plasmid MAXI kit protocol or by $CsCl_2$ isopycnic centrifugation [Ausubel et al., 1995b].
- COS-7 (COS) cells: COS cells need to be split 1 : 3 twice a week in order to keep them in exponential growth. A confluent 75-cm^2 flask should yield ~6 × 10^6 cells and this should be diluted to 1 × 10^5 cells/ml on reseeding.
- Growth medium: DMEM supplemented with 10% FBS, 50 U/ml of penicillin, and 50 μg/ml of streptomycin.
- PBSA
- Trypsin/EDTA: 2.5 mg/ml crude trypsin, 1 mM EDTA in PBSA
- Hemocytometer or electronic cell counter
- DMEM/NS: DMEM, 10% NU-SERUM
- Polypropylene Falcon 2059 tubes, 15 ml
- Transfection cocktail
- Solution 3

Protocol

(a) For each 75-cm^2 flask of exponentially growing COS cells, aspirate the growth medium and wash once with 10 ml of PBSA at 37°C.

(b) Add 2 ml of trypsin/EDTA and rock the flask to distribute the solution evenly. Incubate for 5–10 min at 37°C until the cells begin to detach.

(c) Inactivate the trypsin with 2 ml of growth medium and mix gently to form an even suspension of cells. As the anti-tryptic activity in reduced serum, or serum-free medium, may be insufficient to quench the proteolytic activity of the trypsin, it is important to use serum-containing medium at this step.

(d) Count an aliquot of cells from the suspension with a hemocytometer. COS cells will reach a density of approximately 6 × 10^6 cells per 75-cm^2 flask (~10^5 cells/cm^2) at confluence. For a 35-mm dish (or 1 well of a 6-well plate), plate out a suspension of 1 × 10^5 cells in 1–5 ml of growth medium (~1 × 10^4 cells/cm^2). This number of cells is enough to obtain reliable results and will not overgrow the dish for 3 days.

(e) Incubate the plates in a humidified 37°C, 5.0% CO_2 incubator overnight.

(f) The following day, exchange the growth medium for DMEM/NS. Incubate the plate overnight as before.

(g) For each well to be transfected, prepare at least 2 ml of the transfection cocktail as described in the materials section.

(h) Using sterile (aerosol resistant) pipette tips, mix 1–5 μg/ml of plasmid DNA with 2 ml of the transfection cocktail in sterile 15-ml polypropylene Falcon 2059 tubes. In our experience, 3 μg/ml of plasmid DNA will result in the maximum transfection/expression efficiency with COS-7 cells.

(i) *Gently,* wash each well of the plate twice with 2 ml of PBSA or DMEM (without serum) at 37°C. Aspirate each well in the plate to remove the remaining PBSA (or DMEM). Do not let the cells dry out. If a large number of wells are to be transfected, stagger the washing and transfection steps to allow the manipulation of a manageable number of plates. Three 6-well plates is a reasonable number to handle at any one step of the procedure.

(j) Add the prepared 2 ml of transfection cocktail + DNA to each well.

(k) Incubate the plates for 2–3 h in a humidified, 5.0% CO_2 incubator at 37°C. For COS cells, incubation for 2.5 h will result in optimal transfection. It is important to monitor the cells for signs of toxicity every half hour or so. If the cells begin to detach or look severely vesiculated, proceed to the next step. It is usual and normal for COS-7 cells to look vesiculated when incubated with DEAE-dextran and chloroquine.

(l) Aspirate the transfection cocktail from each well and add 1–2 ml of solution 3 for *EXACTLY 1.5 min.* After 2 min of incubation, the cells will begin to detach and can be lost from the plate in subsequent washing steps.

(m) *Very gently* add, in a dropwise fashion, along the side of each transfected well, one wash of 5 ml of DMEM at *ROOM TEMPERATURE* without serum or antibiotics. This step removes the residual DMSO, which can harm the cells and may affect the experimental results.

(n) Gently add, in a dropwise fashion, along the side of each transfected well, 5 ml of growth medium. Growth for 1 day in this medium allows the cells to recover from the shock induced by the transfection procedure.

(o) Incubate in a humidified, 5.0% CO_2 incubator at 37°C overnight.

(p) In 24 h (and every 24 h thereafter for the next 3 days), exchange the old growth medium for fresh. (If secreted, or accumulated protein is desired; then consult the "collection of secreted proteins" section below).

(q) At each media exchange, or after the desired time has elapsed, harvest the cells or the supernatant for analysis. The time of maximum expression will vary from one cell type to the next. For COS-7 cells, the time of maximum protein expression is between day 2 and day 4 after transfection.

3.2. Abbreviated Procedure for Adherent Cells

Protocol 3.2

The time between plating COS cells and obtaining the results from transfections can be accelerated by 24 h. A modification of Protocol 3.1 can be used with COS cells with good results. Follow Protocol 3.1 and substitute the following where indicated:

(d) Plate 250,000 COS cells per well of a 6-well plate.

(e) Incubate the cells in growth medium for 1–2 h in a 5.0% CO_2 incubator at 37°C.

(f) Replace the growth medium in each well with 2–5 ml of DMEM/NS and incubate overnight in a 5.0% CO_2 incubator at 37°C.

(g) Continue on with Protocol 3.1 at step (g).

Figure 10.1 illustrates the typical results one can expect using Protocol 3.1 to transfect COS cells with the β-galactosidase expression plasmid pSVβGAL. In this titration experiment, the only variable was the concentration of DNA used for each well of the 6-well plate as noted. The results indicate that 3 μg/ml of pSVβGAL is the amount of DNA desired for maximal transfection, expression, and cell viability.

3.3. Collection of Secreted Proteins

Secreted proteins can be collected as they are produced. Days 2–4 post-transfection correspond to the maximum expression and secretion of the product of transfected cDNAs. Secreted proteins can be produced in relatively large amounts; 10,000–100,000 times the amount found in the tissue or cell of origin. These proteins can be collected and purified from the cell supernatants. Depending on the protein, purification to homogeneity can be complicated by the presence of serum proteins contained in the FBS component of tissue culture growth media (Lévesque, 1991). DMEM + 2% NU-SERUM is an easily prepared reduced-serum growth medium for the collection and purification of proteins. The amount of protein in this growth medium is approximately 180 μg/ml, which is considerably less than the 50–80 mg/ml of protein in DMEM

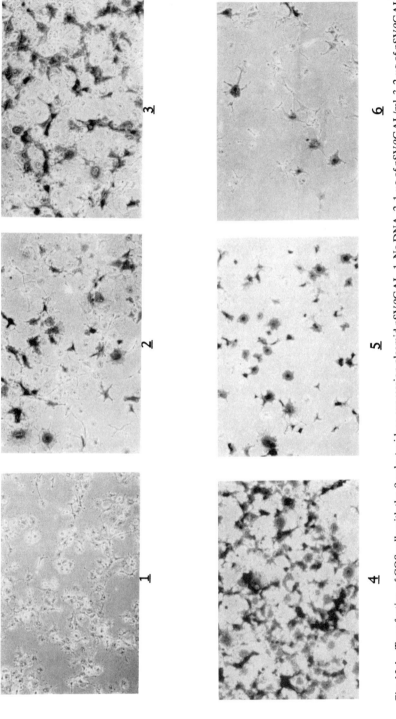

Fig. 10.1 Transfection of COS cells with the β-galactosidase expression plasmid pSVβGAL. 1. No DNA. 2. 1 μg of pSVβGAL/ml. 3. 2 μg of pSVβGAL/ml. 4. 3 μg of pSVβGAL/ml. 5. 4 μg of pSVβGAL/ml. 6. 5 μg of pSVβGAL/ml. Reporter gene expression (β-galactosidase) was determined as described in Protocol 3.5.

containing 10% FBS. If an even lower concentration of contaminating serum proteins is required (e.g., 20 μg/ml), then the reader is referred to the formulation used for Mouse Lta tk$^-$ cells described by Liu et al. [1996]. COS cells can tolerate exposure to either of these low-protein media for 3–4 days and secrete protein at levels comparable to DMEM + 2% NU-SERUM.

3.4. Transfection of Cells in Suspension

Fujita et al. [1986] successfully used a modification of the DEAE-dextran technique to transfect non-adherent lymphoid cells and trypsinized suspensions of adherent cells. This protocol is adapted from Current Protocols in Molecular Biology (Wiley & Sons, Inc.).

Protocol 3.4

Reagents and Materials
Aseptically prepared
- Suspension TBS (STBS)
- DEAE-dextran, 10 mg/ml in STBS
- DNA for transfection
- Trypsin, 0.25% in 1 mM EDTA (to release adherent cells for transfection in suspension)
- Growth medium (DMEM + 10% FBS)
- Hemocytometer or electronic cell counter
- Polypropylene centrifuge tube, 50 ml
- Dimethylsulfoxide (DMSO)

Note: Use sterile, aerosol-resistant pipette tips to transfer DNA into sterile microcentrifuge tubes, using 10 μg of DNA per 2 × 10^7 cells. This amount of DNA is a good starting point for protocol optimization. However, this parameter needs to be adjusted for each cell type to yield acceptable results.

Protocol
(a) Dissolve each DNA to be transfected in 0.5 ml of sterile 1× STBS.
(b) Trypsinize exponentially growing adherent cells for 5–10 min. Neutralize the trypsin by adding an equal volume of growth medium containing 10% FBS. If the cells are already growing exponentially in suspension, skip the trypsinization step and proceed to step (d) directly.
(c) Count an aliquot of the cell suspension with a hemocytometer or electronic cell counter.
(d) Collect 2 × 10^7 cells in a sterile 50-ml polypropylene centrifuge tube by centrifuging the tube at 100–150 g for 5 min at room temperature.

(e) Aseptically aspirate the supernatant, wash the cells in 5 ml of warm STBS solution, and centrifuge as in step (d).

(f) Prepare a solution containing twice the final concentration of DEAE-dextran in STBS. Add 0.5 ml of this solution to the 0.5 ml of DNA solution from Step (a). The amount of DNA, the duration of incubation, and the concentration of DEAE-dextran need to be titrated for each cell type. Final DEAE-dextran concentrations in the range of 100–500 μg/ml will need to be tested in order to ensure maximal levels of transfection. The 10 mg/ml DEAE-dextran stock solution is convenient for dilution to a variety of final concentrations.

(g) Resuspend the cells in 1 ml of the DNA/DEAE-dextran solution and incubate for 30–90 min in a tissue culture incubator at 37°C. Gently swirl the cell transfection cocktail every 30 min to avoid clumping.

(h) Gently stir the cells while adding DMSO dropwise until the final DMSO concentration is 10%. It is important to mix the solution completely.

(i) Incubate the cell suspension with DMSO at room temperature for 2–3 min.

(j) Add 15 ml of STBS to the cells. The addition of the STBS will dilute the DMSO so that the DMSO treatment does not proceed for more than 3 min.

(k) Collect the cells by centrifuging at 150 g for 5 min.

(l) Aspirate the transfection mixture supernatant.

(m) Resuspend and wash the cells in 10 ml of STBS and collect the cells as in Step (d).

(n) Wash the cells with 10 ml of growth medium without serum.

(o) Resuspend the cell pellet in complete growth medium with serum at a density of approximately 1×10^5 to 1×10^6 cells/ml. The cell concentration in the growth medium should be adjusted to allow the cells to grow exponentially for about 48 h, or until the cells are to be harvested and analyzed.

3.5. β-Galactosidase Staining

Protocol 3.5

Reagents and Materials
- COS cells
- 6-well plate
- DNA for transfection
- Growth medium
- Fixation solution
- Staining solution

Protocol

(a) Transfect COS cells according to Protocol 3.1.

(b) Plate 1×10^5 cells in each well of a 6-well plate.

(c) Transfect each well with an increasing amount of DNA/ml as shown in Fig. 10.1.

(d) Allow the cells to recover for 2–3 days in growth medium (with FBS or Nu-Serum).

(e) Assay the cells for β-galactosidase:

   (i)   Wash the cells once with 5 ml of $1 \times$ PBS at *room temperature.*

   (ii) Add 1–2 ml of fixation solution at *room temperature.*

   (iii) Incubate the cells at room temperature in a fume hood for 15 min. Do not allow the cells to fix for more than the indicated time, as this will decrease the sensitivity of the assay.

   (iv) Aspirate the fixation solution and dispose of it properly.

   (v)   Wash the cells twice in 5 ml of PBS at room temperature.

   (vi) Aspirate the wash solution and add at least 1 ml of the staining solution. Incubate the staining cells at 37°C in a tissue culture or humidified incubator for 1 h to overnight, depending on the level of β-galactosidase activity and hence the ability to identify blue cells.

   Note: The cells should start to appear blue after 1 h.

4. NOTES ON SAFETY

The American Type Culture Collection (ATCC) advises that COS cells should be handled according to generally accepted BL-2 containment procedures. It is important to consult with your own biological safety committee and ask them what procedures and controls they require when using this cell line. Information on national guidelines is available from NIH, Environmental Protection Branch, Bldg 13, Room 2W64, NIH, Bethesda, MD; tel: 202-496-7990; internet at <http://www.nci.nih.gov/intra/resource/biosafe.htm>.

Potassium ferro/ferricyanide solutions need to be kept at pH >7.0 to insure that free cyanide gas is not liberated from the solution, as this can occur at a pH below 7.

Glutaraldehyde is a suspected carcinogen and, furthermore, can permanently injure lung tissue. All handling of glutaraldehyde-containing solutions should be performed in a fume hood, and all chemical wastes should be discarded as toxic waste.

5. APPLICATONS

One focus of our lab is the identification and understanding of the enzymes in the heparan sulfate biosynthetic pathway. Heparan sulfate D-glucosaminyl N-deacetylase/N-sulfotransferase type I and II [Orellana et al., 1994] were the first enzymes of this pathway to be constructed with a protein-A tag, yielding transfected COS cells that could produce protein for the determination of enzyme function.

We have also generated wild-type and protein-A expression constructs of our own. Transfection of COS-7 with either the recently cloned wild type [Liu et al., 1996], or protein A heparan sulfate D-glucosaminyl 3-O-sulfotransferase, has been an easy way to produce and purify sufficient quantities of this rare enzyme for detailed analysis of its function *in-vitro*. We have used Protocol 3.1 on both a small (6-well plate) and a large (75-cm^2 flasks) scale. The yield of active, secreted protein (on an activity per cell basis) is on the order of 1×10^5 times the activity secreted by wild-type Lta cell line.

It is important to note that cDNAs with unknown function can be expressed in COS cells and often yield enough active protein to determine their novel activities. Expression cloning of novel cDNAs is the most common instance of this use of the DEAE-dextran transfection into COS cells today, because of the need for small amounts of DNA, and because of the ability to transfect, at once, a large number of cultured cells in a relatively short time.

REFERENCES

Ausubel F, Brent R, Kingston R, Moore D, Seidman J, Smith J, Struhl K (1995a): "Current Protocols In Molecular Biology." New York: John Wiley & Sons, pp 9.2.1–9.2.6.

Ausubel F, Brent R, Kingston R, Moore D, Seidman J, Smith J, Struhl K (1995a): "Current Protocols In Molecular Biology." New York: John Wiley & Sons, pp 1.7.6–1.7.8.

Corsaro C, Pearson M (1981): Enhancing the efficiency of DNA-mediated gene transfer in mammalian cells. Somat Cell Mol Genet 7: 603–616.

Felgner P, Rhodes (1991): Gene theraputics. Nature 349: 351–352

Fujita T, Shibuya H, Ohashi T, Yamanishi K, Taniguchi T (1986): Regulation of human interlukin-2 gene: functional DNA sequences in the 5′ flanking region for the gene expression in activated T lymphocytes. Cell 46: 401–407.

Holter W, Fordis M, Howard B (1989): Efficient gene transfer by sequential treatment of mammalian cells with DEAE-dextran and deoxyribonucleic acid. Exp Cell Res 184: 546–551.

Kahn ML, Lee S, Dichek D (1992): Optimization of retroviral vector-mediated gene transfer into endothelial cells *in vitro*. Circ Res 71: 1508–1517.

Lévesque J-P, Sansilvestri P, Hatzfeld A, Hatzfeld J (1991): DNA transfection in COS cells: a low-cost serum-free method compared to lipofection. Biotechniques 11: 313–318.

Liu J, Shworak N, Fritze L, Edelberg J, Rosenberg R (1996): Purification of heparan sulfate D-glucosaminyl 3-*O*-sulfotransferase. J Biol Chem 271: 27072–27082.

Luthman H, Magnusson (1983): High efficiency polyoma DNA transfection of chloroquine treated cells. Nucl Acids Res 11: 1295–1308.

Maniatis T, Fritsch E, Sambrook. (1989): "Molecular Cloning: A Laboratory Manual." Cold Spring Harbor Press, NY: Cold Spring Harbor Laboratory, pp 16.41–16.81.

McCutchen J, Pagano J (1968): Enhancement of the infectivity of simian virus 40 deoxyribonucleic acid with diethyl-aminoethyl-dextran. J Natl Cancer Inst 41: 351–357.

Orellana A, Hirschberg C, Wei Z, Sweidler S, Ishihara M (1994): Molecular cloning and expression of a glycosaminoglycan *N*-acetylglucosaminyl *N*-deacetylase/*N*-sulfotransferase from a heparin producing cell line. J Biol Chem 269: 2270–2276.

Pearson A, Lux A, Krieger M (1995): Expression cloning of dSR-CI, a Class C macrophage-specific scavenger receptor from *Drosophila melanogaster*. Proc Natl Acad Sci USA 92: 4056–4060.

Puchalski R, Fahl W (1992): Gene transfer by electroporation, lipofection, and DEAE-dextran transfection: compatability with cell-sorting by flow cytometry. Cytometry 13: 23–30.

Reeves R, Gorman C, Howard B (1985): Minichromosome assembly of nonintegrated plasmid DNA transfected into mammalian cells. Nucleic Acids Res 13: 3599–3615.

Schwachtgen J-L, Ferreira V, Meyer D, Kerbiriou-Nabias D (1994): Optimization of the transfection of human endothelial cells by electroporation. Biotechniques 17: 882–887.

Seldon R, Burke K, Rowe M, Goodman H, Moore D (1986): Human growth hormone as a reporter gene in regulation studies employing transient gene expression. Mol Cell Biol 6: 3173–3179.

Sussman D, Milman G (1984): Short-term, high-efficiency expression of transfected DNA. Mol Cell Biol 4: 1641–1643.

APPENDIX: MATERIALS AND SUPPLIERS

Material	Catalogue Number	Supplier
Chloroquine diphosphate	C-6628	Sigma
COS-7 cells	CRL1651	A.T.C.C.
DEAE-dextran (MW 500,000)	17-0350-01	Pharmacia Biotech
DMEM (high glucose)	12100-061	Life Technologies Inc.
DMSO (tissue culture qualified)	D-2650	Sigma
FBS	12103-78P	JRH Biosciences Inc.
Flasks, 75 cm²	153789	Nalge-Nunc
Flasks, 175 cm²	156502	Nalge-Nunc
Glutaraldehyde, 2.5 M	49630	Fluka Chemical Corp.
Hemocytometer	15170-079	V.W.R.
KH_2PO_4	P-5379	Sigma
K_2HPO_4	P-5504	Sigma
N,N-dimethylformamide (DMF)	D-4254	Sigma
Nu-Serum	55000	Becton Dickinson Labware
Penicillin G (10,000 U/ml), streptomycin sulfate (10,000 μg/ml)	15140-122	Life Technologies Inc.
Phosphate buffered saline	14040-133	Life Technologies Inc.

APPENDIX (Continued)

Pipette tip P20	53502-903	V.W.R. Scientific
Pipette tip P200	53502-901	V.W.R. Scientific
Pipette tip P1000	53502-905	V.W.R. Scientific
Plasmid: pSVβGAL	E1081	Promega Corporation
Plates, 6-well, 35-mm diameter well size	3506	Corning Costar
Polypropylene centrifuge tubes, 14 ml, Falcon	4-2059-1	Becton Dickinson
Polypropylene centfrifuge tubes, 6 ml, Falcon	4-2063-1	Becton Dickinson
Potassium ferricyanide, $K_3Fe(CN)_6$	P-8131	Sigma
Potassium ferrocyanide, $K_4Fe(CN)_6 \cdot 3H_2O$	P-9387	Sigma
Qiagen Maxi Kit (10)	12162	Qiagen Inc.
Trypsin, EDTA: 0.25% trypsin, 1.0 mM EDTA	25200-072	Life Technologies Inc.
X-Gal, 100 mg	15520-034	Life Technologies Inc.

Cationic Liposomes

Vadim V. Bichko

Scriptgen Pharmaceuticals, Inc., Medford, MA

DNA Transfer to Cultured Cells, Edited by Katya Ravid and R. Ian Freshney.
ISBN 0-471-16572-7   © 1998 Wiley-Liss, Inc.

I. INTRODUCTION

Cationic lipid-mediated DNA transfection into cultured cells was first reported by Felgner and coauthors in 1987 [Felgner et al., 1987]. A major improvement to the method was achieved in 1993 by replacement of the monocationic lipid reagent with a polycationic one, Lipofectamine [Hawley-Nelson et al., 1993].

The method is based upon an ionic interaction of DNA and liposomes to form a complex, which can deliver functional DNA into cultured cells. The interaction of DNA and charged lipids can give rise to a variety of structures. Plasmid DNA is complexed, but not encapsulated within unilamellar liposomes, 600–1200 nm in size [Mahato et al., 1995b], formed by cationic lipids in water. Bilayer-covered DNA tubules have been also observed, using freeze-fracture electron microscopy, [Sternberg et al., 1994]. An excess of charged lipids leads to formation of multilamellar structures, where DNA is trapped between the lamellae [Gustafsson et al., 1995].

The precise mechanism of DNA delivery into the cell has not yet been determined. It was initially proposed that DNA entry is mediated by adsorption of liposomes to, and fusion with, the cellular membrane. However, recent studies, based on electron microscopy observations, suggest that the main route of entry of cationic liposomes into cells is by endocytosis [Friend et al., 1996]. In the same study, occasional entry of liposomes into the nucleus by fusion with the nuclear envelope was observed. Upon entering the cell by endocytosis, DNA is released into the cytoplasm by destabilization of endosomal membrane [Xu et al., 1996].

The physiological effects of liposome-mediated transfection are yet to be determined. One report indicates that, during transfection of keratinocytes with Lipofectamine, there is release of a substantial amount of interleukin-1 α (IL-1α), which could affect the evaluation of the results of expression of some genes [Komine et al., 1994]. Another report showed that it was possibile to separate the transfecting and cytotoxic activities of the transfection mixture by purification of DNA-liposome complexes from uncomplexed liposomes [Hofland et al., 1996].

The advantages of cationic liposome-mediated transfection over other methods include generally higher efficiency, the ability to transfect successfully a wide variety (over 300 reported) of eucaryotic cell lines, many of which are refractive to other transfection procedures, and relatively low cell toxicity. Another advantage is that the basic procedure of DNA transfection can be adapted for transfection with RNA, synthetic oligonucleotides, proteins, and viruses. Finally, cationic liposomes can be used for the successful delivery of functional genes or viral genomes *in vivo* [Mahato et al., 1995a; Tagawa et al., 1996; Thorsell et al., 1996].

The disadvantage of liposome reagents is the relatively high cost, which practically precludes their use for large-scale (several 100-mm plates or more) transfections. However, it should be mentioned that cationic liposome reagents suitable for transfections, at least with DNA, can be prepared by a simple ethanol injection method using commercially available and inexpensive cationic and neutral lipids [Campbell, 1995]. According to reporter gene (luciferase) activity, the efficiency of transfection was practically the same for Lipofectamine and liposomes prepared by ethanol injection.

In this paper we will provide some basic protocols for DNA transfections; however, we will focus on novel applications of cationic liposomes, such as delivery of virions and viral components.

2. REAGENTS

2.1. Reagents for Cell Transfections

Cationic lipids

Lipofectamine
3:1 (w/w) liposome formulation of the polycationic lipid DOSPA (2,3-dioleyloxy-N[2(sperminecarboxamido)ethyl]-N,N-dimethyl-1-propanaminium trifluoroacetate) and the neutral lipid DOPE (dioleoylphosphatidylethanolamine) in water.

Lipofectin
1:1 (w/w) liposome formulation of the cationic lipid DOTMA (N-[1-(2,3 dioleyloxy)-propyl]-n,n,n-trimethylammonium-chloride) and the neutral lipid, DOPE, in water.

LipofectACE
1:2.5 (w/w) liposome formulation of the cationic lipid DDAB (dimethyl dioctadecylammonium bromide) and a neutral lipid, DOPE, in water.

Complete growth medium
DMEM, 1×, liquid
10% FBS
Penicillin, 100 U/ml
Streptomycin, 100 μg/ml

Reduced serum medium
OPTI-MEM 1 (contains 2% FBS)

Serum-free medium
DMEM without serum or antibiotics

PBS
NaCl	0.15	M
Na_2HPO_4	0.01	M
KH_2PO_4	0.002	M
KCl	0.003	M

PBSA
NaCl	0.15	M
K_2HPO_4	0.006	M
KH_2PO_4	0.002	M

Trypsin
Trypsin, 0.25%, in PBS lacking Ca^{2+} and Mg^{2+} (PBSA)

2.2. Reagents for CAT Assays

Buffers
Tris HCl, 1 M(pH 8.0).
Tris HCl, 0.1 M(pH 8.0).
Tris HCl, 0.1 M(pH 8.0), 0.1% Triton X-100. Store at 4°C.

Substrate
250 mM chloramphenicol (in 100% ethanol). Store in aliquots at −70°C.

CAT enzyme standards
Prepare standard solutions of 0.2, 1, 2, 4, and 10 U/ml in CAT dilution Buffer (0.1 M Tris-HCl [pH 8.0], 50% glycerol, 0.2% BSA). Store at 4°C.

Liquid scintillation cocktail
Econofluor™ (BRL)

Isotope
[^{14}C]butyryl coenzyme A (0.010 mCi/ml)

2.3. Reagents for β-gal Assays

Substrate
X-gal, 20 mg/ml in dimethylformamide
    Store at −20°C in the dark, in a polypropylene tube, for up to 6 months.

Fixative
Formaldehyde 1.8%
Glutaraldehyde 0.05%
in PBSA

Prepare by mixing 85 ml of water, 10 ml 10× PBSA, 5 ml formalin (37% formaldehyde solution), 0.2 ml glutaraldehyde (25% solution). Store at 4°C.

Stain solution
5 mM potassium ferricyanide, 5 mM potassium ferrocyanide, 2 mM $MgCl_2$ in PBSA. Store at 4°C.

Substrate/stain solution
1 mg/ml X-gal in stain solution. Prepare immediately before using.

3. PROTOCOLS

3.1. Transfection of Adherent Cells with DNA

Plate cells on 35-mm dishes so that they are 50–90% confluent on the day of transfection.

Protocol 3.1

Reagents and Materials
Aseptically prepared
- Cells for transfection
- Cationic lipid solution: Lipofectamine, Lipofectine, or LipofectACE
- Complete growth medium
- Reduced serum medium
- Serum-free medium
- PBSA
- 0.25% Trypsin, Ca^{2+} and Mg^{2+} free.
- Multiwell plates, 6-well 35-mm

Protocol
(a) In 6-well 35-mm dishes, seed approximately 1×10^5 to 3×10^5 cells per well in 3 ml of growth medium appropriate to the cells being used.
(b) Incubate the cells at 37°C in a CO_2 incubator until the cells are 50–90% confluent. This will usually take 18–24 h, but should not take less then 16 h.
(c) Before transfection, prepare the DNA and lipid solutions in sterile tubes. For the DNA solution, dilute 1–2 μg of DNA in 0.5 ml

of reduced serum medium. For the lipid solution, dilute 2–25 μl of cationic lipid reagent in 0.5 ml of serum-free medium. Combine the two solutions, mix gently, and incubate at room temperature for 15–45 min to allow the formation of DNA-lipid complexes.

(d) Rinse the cells once with 2 ml of serum-free medium.

(e) Overlay the complex solution onto the cells. Antibacterial agents should be omitted during transfection.

(f) Incubate the cells with the complexes at 37°C in a CO_2 incubator for 2–24 h. Usually, a period of 5 or 6 h is enough.

(g) Following incubation, remove the transfection mixture from the cells and replace it with 3 ml of complete growth medium.

(h) Replace the growth medium 18–24 h after the start of transfection.

(i) Assay the growth medium or cells for transient gene activity as appropriate (see Section 3.5 or 3.6 for examples).

3.2. Transient Transfection of Suspension Cells with DNA

Protocol 3.2

Reagents and Materials
Aseptically prepared
- Complete growth medium
- Cationic lipid solutions
- Reduced serum medium
- Serum-free medium
- Petri dishes, 35-mm

Protocol

(a) Prepare transfection mixture in sterile tubes as follows:
Dilute 2–5 μg of DNA in 0.5 ml of reduced serum medium; then dilute 2 to 20 μl of cationic lipid reagent in 0.5 ml of serum-free medium.

(b) Combine the two solutions, mix gently, and incubate at room temperature for 15–45 min to allow the formation of DNA-lipid complexes.

(c) Centrifuge a cell suspension containing approximately 1×10^6 to 2×10^6 cells, and aspirate the medium. Resuspend the cells in transfection mixture and transfer to a 35-mm dish.

(d) Incubate the cells in a CO_2 incubator for 4–6 h.

(e) To each dish, add 0.5 ml of growth medium, appropriate for the cells being used, supplemented with 30% serum (high serum to protect cells in suspension, which are more sensitive to toxic effects of liposome reagents).

(f) Incubate in a CO_2 incubator overnight.

(g) To each dish, add 2 ml of complete growth medium and incubate in CO_2 incubator.

(h) At 24–72 h after the start of transfection, assay the cells or medium for gene activity as appropriate.

3.3. Transfection of Adherent Cells with RNA

Lipofectin reagent should be used in this protocol. For a successful transfection, mRNA should be capped and polyadenylated, and also should contain 5′-untranslated sequences [Malone, 1989].

Protocol 3.3

Reagents and Materials

Aseptically prepared
- Cells for transfection
- Complete growth medium
- Reduced serum medium
- Lipofectamine
- RNA solution for transfection
- Petri dishes, 35-mm

Nonsterile equipment
- Vortex mixer

Protocol

(a) Plate approximately 2×10^5 to 3×10^5 cells in 35-mm dishes in 2–3 ml of complete growth medium in order to achieve 70–90% cell confluency the next day (day of transfection).

(b) Wash the cells once with reduced serum medium at room temperature.

(c) For each transfection, prepare RNA-Lipofectin complexes as follows:
   (i) Add 5–15 μl of Lipofectamine reagent to 1 ml of reduced serum medium and vortex briefly.
   (ii) Add 2.5–5 μg RNA and vortex briefly.

(d) Immediately add the prepared RNA-Lipofectin complexes to washed cells.

(e) Incubate at 37°C in a CO_2 incubator for 2–24h, then replace the transfection mixture with complete growth medium.

(f) At 16–24h after the start of transfection, assay for the gene activity as appropriate.

3.4. Transfection with Proteins

Protocol 3.4

This protocol was used by Sells and coworkers [Sells et al., 1995] for Lipofectamine-mediated delivery to cultured cells of several biologically active proteins, including β-galactosidase and Ras protein. Other cationic liposome reagents, such as Lipofectin and LipofectACE, were also used for transfections with proteins, generally with lower efficiency [Debs et al., 1990; Baubonis and Sauer 1993; Gao and Huang 1993; Lin et al., 1993]. Up to 50% of cells were found to be transfected, as determined by *in situ* β-galactosidase assay [Sells M. A. and Chernoff J., personal communication]

Reagents and Materials
Aseptically prepared
- Complete growth medium for plating
- PBSA
- Cells for transfection
- Protein for transfection
- Serum-free, antibiotic-free DMEM
- Lipofectamine
- Petri dishes, 35-mm

Protocol
(a) Adjust purified proteins to 1 mg/ml in PBSA.
(b) Plate 10^5 cells in 35-mm dishes 1 day prior to transfection.
(c) On the day of transfection, prepare the transfection mixture as follows:
   (i) Dilute 50 μg of protein in 0.1 ml of serum-free, antibiotic-free DMEM.
   (ii) Dilute 10 μg (5 μl) of Lipofectamine in 0.1 ml of the same medium.
(d) Combine the two solutions, mix, and incubate at room temperature for 20 min.
(e) Remove the medium, dilute the transfection mix with 0.8 ml of serum-free DMEM, and add to the cells.
(f) Incubate the cells at 37°C in a CO_2 incubator for 2 h.
(g) Harvest the cells and assay for protein activity.

3.5. Chloramphenicol Acetyltransferase (CAT) Assay

Protocol 3.5

An expression plasmid for CAT (for example, pCMVCAT [Boshart et al., 1985]) can be used as a reporter gene for transfection.

Reagents and Materials
Aseptically prepared
- PBSA
- Multiwell plates: 6-well, 35-mm
Nonsterile
- Tris buffer

- Tris/Triton X-100
- Chloramphenicol, 250 mM in 100% ethanol
- CAT standard solutions
- [^{14}C]CoA: [^{14}C]butyryl coenzyme A, 10 μCi/ml
- DDW: deionised, distilled water
- Econofluor™
- Polypropylene scintillation vials, 3.5 ml
- Microcentrifuge tubes
- Microcentrifuge
- Ice bath

Protocol

Cell harvesting for 6-well, 35-mm, plates
(a) At 24–72 h after transfection, wash the cells once with PBSA.
(b) Put the plates on ice and add per each well 1 ml of Tris/Triton.
(c) Freeze for 2 h at −70°C.
(d) Thaw plates at 37°C, then put on ice.
(e) Transfer the cell lysates to microcentrifuge tubes and spin for 5 min at maximum speed.
(f) Collect the supernatants and heat at 65°C for 10 min to inactivate inhibitors of CAT.
(g) Centrifuge at maximum speed for 3 min and collect the supernatant ("cell extract"). Keep at −70°C.

CAT assay [Based on Neumann et al., 1987, modified by Life Technologies].

(a) Put 5–150 μl of "cell extract" from each sample into a 3.5-ml polypropylene scintillation vial and add 0.1 M Tris to a final volume of 150 μl.
(b) For a negative control, use 150 μl of 0.1 M Tris.
(c) For a positive control, add 150 μl of 0.1 M Tris to each of 5 vials.
(d) Add 5 μl of each CAT standard solution to give a standard curve of 1, 5, 10, 20 and 50 mU of CAT.
(e) To each sample (including controls), add 100 μl of the following mixture:
    (i)   84 μl DDW
    (ii) 10 μl of 0.1 M Tris
    (iii) 1 μl of chloramphenicol
    (iv) 5 μl (50 nCi) of [^{14}C]-CoA
(f) Incubate capped samples at 37°C for 2 h.
(g) Add 3 ml of Econofluor™ to all tubes and cap. Mix by inverting the tubes.

(h) Incubate at room temperature for 2 h.

(i) Count the samples for 0.5 min in a liquid scintillation counter.

3.6. *In Situ* Staining for β-Galactosidase (β-gal) with X-gal

Protocol 3.6

[Based on Sanes et al., 1986, modified by Life Technologies]

In this case expression plasmid for β-gal would be used as a reporter gene for transfections (for example, pCMVβgal [MacGregor and Caskey, 1989]).

Reagents and Materials

Nonsterile

- PBS
- PBSA
- Fixative (see Section 2.3)
- Substrate/stain solution (see Section 2.3)
- 10% formalin in PBSA

Protocol

(a) Wash the cells once with 2 ml of PBS.

(b) Fix the cells with 1 ml of fixative for 5 min at room temperature.

(c) Wash the cells twice with 2 ml of PBSA.

(d) Add 1 ml per well of substrate/stain solution and incubate 2 h to overnight at 37°C.

(e) Rinse the cells in each well with 2 ml of PBSA. Observe on an inverted microscope, and count the blue (β-gal positive) cells.

(f) To store plates, fix the cells in each well with 1 ml of 10% formalin in PBSA for 10 min at room temperature, rinse with PBSA, and store in PBSA at 4°C.

4. TRANSFECTION OF CULTURED CELLS WITH HEPATITIS DELTA VIRUS AND VIRAL COMPONENTS

As was previously established, cationic liposome reagents can facilitate the infection of cultured cells with some viruses [Konopka et al., 1990; Konopka et al., 1991]. Furthermore, virus can be efficiently introduced into cultured cells, which cannot be infected in the absence of liposome reagents [Faller and Baltimore 1984; Innes et al., 1990; Bass et al., 1992]. As we have reported previously, hepatitis delta virus (HDV) can be introduced into several cultured cell lines using Lipofectamine [Bichko et al., 1994]. Thus, the cationic lipid DNA transfection procedure was successfully extended to the delivery of the intact viral particles. Viral RNA genomes can also initiate viral replication upon delivery to cultured cells via cationic liposome transfection. Such results were obtained with reovirus genomic RNA [Roner et al., 1990], and duck hepatitis B virus pregenomic RNA [Huang and Summers, 1991]. In this study, we tried to extend such experiments to a negative-stranded RNA virus, hepatitis delta virus (HDV), in order to deliver different components of virions and determine the minimum requirements for initiation of HDV replication and expression in cultured cells.

4.1. Transfections with Virions

All the transfection experiments described in this study were performed with Lipofectamine. HDV was obtained from the sera of acutely infected woodchucks. Virions were collected by two rounds of centrifugation through layers of 10% and 20% sucrose, as described [Bichko et al., 1996b]. The human hepatoma cell line Huh7 [Nakabayashi et al., 1982] was used for optimization of transfection.

The transfection was performed basically as for DNA (see Protocol 3.1.), but a preparation of virions was used. Cells were plated on 16-mm tissue culture wells with glass coverslips and were transfected the next day at approximately 50% confluence [Bichko et al., 1994]. To prepare the transfection mixture, 2 μl of a concentrated virion preparation and variable amounts of Lipofectamine (0–5 μl) were each diluted in 125 μl of Opti-MEM, mixed gently, and combined. Following incubation for 30 min at room temperature, the transfection mixture was added to the cells. After incubation for 4 h, the transfection mixture was replaced with DMEM supplemented with 10% fetal calf serum.

Efficient replication of HDV in transfected cells was demonstrated at 5 or 6 days after transfection by Northern analyses [Wu et al., 1995] of HDV for both genomic and antigenomic RNA. Since antigenomic RNA was not present in the virions used for transfection, detection of these species is the most convincing proof of HDV genome replication. However, this method did not allow us to determine which fractions of the cells were positive for HDV. This was achieved with indirect immunofluorescence microscopy to detect the major protein encoded by the HDV genome, delta antigen, or δAg. Cells were doubly labeled: with DAPI, to visualize the nuclei of all cells, and with antibodies to δAg, to detect cells replicating HDV. The results of these experiments are shown in Fig. 11.1. Panels A and B show the same field of cells stained with DAPI (Panel A), and with anti-δAg antibodies, to detect HDV-positive cells (Panel B). A colony of HDV-positive cells can be seen in Panel B.

Fig. 11.1. Immunofluorescence of Huh7 cells transfected with HDV and its RNP. Cells were immunostained to detect δAg (B, D, F, and H), and also stained with DAPI, to visualize nuclei (A, C, E, and G). Transfections were as follows: HDV (A–D); vRNP (E and F); nRNP (G and H), Cells were stained at 6 days after transfection, with the exception of panels C and D (18 days). The bar indicates the scale. The images were acquired with a CCD camera and processed with Adobe Photoshop and Canvas software.

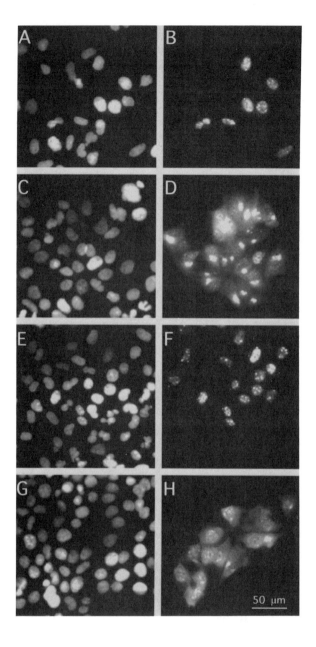

Using this approach, optimization of the Lipofectamine concentration in the transfection mixture was performed. As shown in Table 11.1, the efficiency of transfection was improved, reaching a maximum at 15%, with increases in Lipofectamine concentration. It should be pointed out that further increases of Lipofectamine concentration (more than 5 μl per 250 μl of transfection mixture) led to considerable cell death. In this respect transfection with

Table 11.1. Efficiency of Transfection of Huh7 Cells with HDV Using Different Amounts of Lipofectamine[a]

Lipofectamine, μl[b]	HDV-positive cells, %
0	0
1	<0.1
2	1
3	3
4	7
5	15

[a] Data representative of four experiments.
[b] For transfection of cells in 16-mm wells (24-well plate).

virions was similar to transfection with proteins [Sells et al., 1995], rather than with DNA, where the optimal concentration of cationic liposome reagent could be easily determined. No HDV-positive cells were found in the absence of Lipofectamine (Table 10.1). In separate large-scale experiments, such cells were found, but they did not exceed 0.005% of the population [V. Bichko and J. Taylor, 1996, unpublished observations]. To exclude the possibility that the input virus rather than replicating viral progeny was detected by immunofluorescence, we performed the analysis at 2 days after transfection, and no positive cells were found (results not shown).

The concentration of virions in these experiments was constant and corresponded to approximately 1.5×10^5 RNA-containing particles per cultured cell at the time of transfection. When we tried to decrease this number, a rapid decline in transfection efficiency was observed; on the other hand, we were also unable to increase this number because of massive cell death. Thus, 10^5 RNA-containing particles per cultured cell were used for further transfection experiments.

Other cell lines were also examined for their ability to be transfected with HDV. As Table 11.2 shows, we were able to transfect all four cell lines tested, but the efficiency of transfection varied considerably. This could be explained by the different ability of cell lines to support HDV replication. However, numerous mammalian cell lines were reported to efficiently replicate HDV [Lai,

Table 11.2. Efficiency of Transfection of Different Cell Lines with HDV

Cell Line	HDV-positive Cells, %	Reference
Huh7	15	Nakabayashi et al., 1982
7777 (rat hepatoma)	13	Morris et al., 1981
BHK-21	6	Karabatsos and Buckley, 1967
MDCK	0.2	Leighton et al., 1969

1995]. On the other hand, similar differences between those cell lines were also observed when we used HDV cDNA instead of virions. Thus, those differences in transfection efficiency were not caused by using virions instead of DNA for transfection or by using cell lines with different abilities to replicate HDV.

4.2. Transfection with HDV RNP

Since cells can be efficiently transfected with virions, we next tested whether ribonuceoprotein (RNP) complexes can be transfected into cells as well. There are three sources of HDV RNP. First, viral RNP, or vRNP, derived from virions by removal of the viral envelope in the presence of NP 40 and DTT [Ryu et al., 1993]. Such vRNPs were able to initiate HDV genome replication after transfection into Huh7 cells using Lipofectamine [Bichko et al., 1994]. Second, nuclear RNP, or nRNP, can be isolated from the nuclei of cells, replicating the HDV genome [Ryu et al., 1993. Third, RNP can be isolated by the removal of viral envelope proteins from viruslike particles, as secreted by transfected cells. These RNPs were considered to be identical or very similar to vRNPs; however, recent data indicate that they are different in size and composition [Dingle, K.E., and Taylor, J.M. 1996, personal communication]. Since the nature of those differences remains unclear, these RNPs were not used for transfection experiments.

Nuclear RNPs (nRNPs) were isolated from Huh7 cells transfected with HDV cDNA as described previously [Ryu et al., 1993]. Transfection was performed as for virions, but using 10^5 RNA-containing particles (nRNPs) per transfection.

The results of these experiments are shown in Fig. 11.1 (Panels E–H). When vRNPs were used for transfection (Panels E and F), the results were very similar to those obtained with virions (Panels A and B). The HDV-positive colonies were detected, and distribution of the δAg (nucleoplasmic with some discrete nuclear dots) was typical for early stages (6 d after transfection) of HDV replication in transfected cells [Bichko and Taylor, 1996]. Transfection with nRNP also led to the initiation of genome replication. The efficiency of transfection was similar to that of vRNP and 2–3 times lower than with virions; however, the intracellular distribution of δAg in such cells was different (Panels G and H). Predominantly cytoplasmic (including Golgi) staining was observed. Such a pattern of δAg distribution is characteristic of the late stages (18 days) of HDV replication [Bichko and Taylor, 1996].

4.3. Transfection with HDV RNA

It was demonstrated previously that *in vitro* transcribed linear HDV RNA can initiate HDV replication after transfection of cultured cells, but only in cells stably producing δAg-S (small form of hepatitis delta antigen) [Glenn et al., 1990]. We have extended these experiments with HDV RNA transfection to circular viral RNA, as extracted from virions, and the same result was obtained [Bichko et al., 1994]. Next we tried to transfect naked RNA isolated from vRNP and nRNP. The same negative result was obtained for Huh7 cells with RNA from vRNP. As a control for the quality of RNA, the same preparation was transfected into Cos-Ag cells (stably producing δAg-S), and up to 60% of HDV-positive cells were obtained at 5 days after transfection. However, when we used RNA from nRNP, we did detect a low, but significant and reproducible fraction of cells, replicating HDV. This fraction reached 0.5% and was about 20 times less than for transfections with a corresponding amount of nRNP. It seemed unlikely, but possible, that HDV replication in those cells was initiated by residual plasmid DNA, since cDNA-transfected cells were used to isolate nRNP and finally to prepare naked RNA. To eliminate this possibility, we treated the RNA preparation with RQ1 DNAse (Promega) or RNAse A (Promega) prior to transfection. We found that DNAse pretreatment has practically no effect on the fraction of HDV-positive cells after transfection, while no positive cells were detected with RNAse pretreatment.

To follow the fate of HDV RNA after Lipofectamine-mediated delivery to the cells, we used synthetic FITC-labelled HDV RNA. This was made in vitro by T7 polymerase (Promega), as recommended by manufacturers, in the presence of UTP conjugated with FITC. After 12 h, excess RNA-Lipofectamine complexes were removed from the growth medium by several washes and medium replacements. FITC-labelled RNA was visible inside cells, as shown in Fig. 11.2. Several observations were made. First,

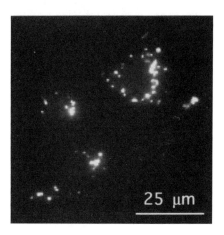

25 μm

Fig. 11.2. Fluorescent detection of Cos-Ag cells at 12 h after transfection with FITC-labelled synthetic HDV RNA. RNA can be seen in cytoplasmic vesicles, concentrated in the perinuclear area of cells. The positions of the nuclei were confirmed by staining with DAPI (not shown). The bar indicates the scale. The image was acquired with a CCD camera and processed with Adobe Photoshop and Canvas software.

FITC-RNA was detected inside practically 100% of Huh7 or Cos-Ag cells. Second, the intensity of the signal decreased rapidly but positive cells could be followed until 5 days after transfection. Third, intracellular FITC-RNA was observed in cytoplasmic vesicles in the perinuclear area, and not in the nucleus, where HDV RNA presumably replicates. Finally, no difference was observed when the cells producing δAg-S were used for transfection.

5. DISCUSSION

Transfection experiments described earlier in this study show that HDV RNP can initiate viral genome replication in transfected cells, while naked HDV RNA cannot. These results are consistent with previous reports showing that both HDV RNA and δAg-S are absolutely required for genome replication in transfected cells [Kuo et al., 1989].

Although both vRNP and nRNP were positive for the initiation of replication, differences were observed in intracellular distribution of δAg. As was reported previously [Bichko et al., 1996a; Bichko and Taylor, 1996], the distribution of δAg in cells replicating HDV changes from nuclear at the early stages (6 days) to cytoplasmic at late stages (18 days), consistent with the accumulation of mutant genomes and consequent expression of altered forms of δAg. Thus, nRNPs isolated from cells at 12 days after transfection contain such mutated genomes. When these nRNPs are introduced into new cells, HDV genome replication is initiated, and mutated genomes continue to accumulate, so that at 6 days after transfection with nRNP (Fig. 10.1, Panel H) the extent of replication and mutation is similar to that observed at 18 days after transfection with virions (Fig. 10.1, Panel D). Taken together, these data suggest that extensive selection for both genomic RNA and δAg occurs during the assembly and secretion of HDV virions.

Another interesting result was obtained with RNA extracted from nRNP. Unlike HDV RNA from any other source, this RNA was able to initiate HDV replication in transfected cells. One possible explanation is that since nRNP, unlike vRNP, contains both genomic and antigenomic RNA, the latter can be used as mRNA to produce sufficient quantities of δAg to initiate replication. Alternatively, δAg could be made initially from its mRNA, a potential contaminant of the nRNP preparation.

Transfections with FITC-labelled HDV RNA enabled us to follow its fate upon cell entry. We observed that this RNA was accumulated in the cytoplasm and did not enter the nucleus at a detectable level. Interestingly, no difference in RNA disposition was observed when cells producing δAg-S were used for transfection. These results do not support the hypothesis that Ag-S mediates nuclear import of HDV RNA. Since this manuscript has been in production, the nuclear import of HDV RNA by SAg was reported by Chou and colleagues (Chou et al., 1998).

Replication of HDV in Cos-Ag cells transfected with FITC-RNA was not confirmed in these experiments because FITC presumably interferes with this process, but replication in such transfected cells is well established. Since HDV RNA was not detected in the nucleus upon transfection, it may be that a minor fraction of the RNA delivered to cell by transfection is actually imported to the nucleus, and this is sufficient to initiate replication.

In conclusion, novel applications of cationic liposomes, such as to the delivery of intact virions, viral ribonucleoprotein complexes, and other viral components, provide powerful tools to study viral replication in cultured cells, especially for negative strand RNA viruses.

ACKNOWLEDGMENTS

All HDV results presented here were obtained in the lab of John Taylor at Fox Chase Cancer Center, Philadelphia, Pennsylvania. We thank John Taylor for the support, Kate Dingle for the sharing of unpublished results and for critical reading of the manuscript, Daphne Bell for CCD camera work, and Dmitri Bichko for digital image processing.

REFERENCES

Bass D, Baylor M, Chen C, Mackow EM, Bremont M, Greenberg HB (1992): Liposome-mediated transfection of intact viral particles reveals that plasma membrane penetration determines permissivity of tissue culture cells to rotavirus. J Clin Invest 90: 2313–2320.

Baubonis W, Sauer B (1993): Genomic targeting with purified Gre recombinase. Nucleic Acids Res 21: 2025–2029.

Bichko VV, Khudyakov Y, Taylor JM (1996a): A novel form of hepatitis delta antigen. J Virol 70: 3248–3251.

Bichko V, Lemon SM, Wang JG, Hwang S, Lai MM, Taylor JM (1996b): Epitopes exposed on hepatitis delta virus ribonucleoproteins. J Virol 70: 5807–5811.

Bichko VV, Taylor JM (1996): Redistribution of the delta antigen in cells replicating the genome of hepatitis delta virus. J Virol 70: 8064–8070.

Bichko V, Netter HJ, Taylor JM (1994): Introduction of hepatitis delta virus into animal cell lines via cationic liposomes. J Virol 68: 5247–5252.

Boshart M, Weber F, Jahn G, Dorsch-Hasler K, Fleckenstein B, Schaffner W (1985): A very strong enhancer is located upstream of an immediate early gene of human cytomegalovirus. Cell 41: 521–530.

Campbell M (1995): Lipofection reagents prepared by a simple ethanol injection technique. BioTechniques 18: 1027–1032.

Chou H-C, Hsieh T-Y, Sheu G-T, Lai MMC (1998): Hepatitis delta antigen mediates the nuclear import of hepatitis delta virus RNA. J Virol 72:3684–3690.

Debs RJ, Freedman LP, Edmunds S, Gaensler KL, Duzgunes N, Yamamoto KR (1990): Regulation of gene expression in vivo by liposome-mediated delivery of purified transcription factor. J Biol Chem 265: 10189–10192.

Faller D, Baltimore D (1984): Liposome encapsulation of retrovirus allows efficient superinfection of resistant cell lines. J Virol 49: 269–272.

Felgner PL, Gadek TR, Holm M, Roman R, Chan HW, Wenz M, Northrop JP, Ringold GM, Danielsen M (1987): Lipofection: a highly efficient, lipid-mediated DNA-transfection procedure. Proc Natl Acad Sci USA 84: 7413–7417.

Friend DS, Papahadjopoulos D, Debs RJ (1996): Endocytosis and intracellular processing accompanying transfection mediated by cationic liposomes. Biochim Biophys Acta 1278: 41–50.

Gao X, Huang L (1993): Cytoplasmic expression of a reporter gene by codelivery of T7 RNA-polymerase and T7 promoter sequence with cationic liposomes. Nucleic Acids Res 21: 2867–2872.

Glenn JS, Taylor JM, White JM (1990): In vitro-synthesized hepatitis delta virus RNA initiates genome replication in cultured cells. J Virol 64: 3104–3107.

Gustafsson J, Arvidson G, Karlsson G, Almgren M (1995): Complexes between cationic liposomes and DNA visualized by cryo-TEM. Biochim Biophys Acta 1235: 305–312.

Hawley-Nelson P, Ciccarone V, Gebeyehu G, Jessee J, Felgner PL (1993): Lipofectamine reagent: a new, higher efficiency polycationic liposome transfection reagent. Focus 15: 73–78.

Hofland H, Shephard L, Sullivan SM (1996): Formation of stable cationic lipid/DNA complexes for gene transfer. Proc Natl Acad Sci USA 93: 7305–7309.

Huang M, Summers J (1991): Infection initiated by the RNA pregenome of a DNA virus. J Virol 65: 5435–5439.

Innes CL, Smith PB, Langenbach R, Tindall KR, Boone LR (1990): Cationic liposomes (Lipofectin) mediate retroviral infection in the absence of specific receptors. J Virol 64: 957–961.

Karabatsos N, Buckley S (1967): Susceptibility of the baby-hamster kidney-cell line (BHK-21) to infection with arboviruses. Am J Trop Med Hyg 16: 99–105.

Komine M, Freedberg IM, Blumenberg M (1994): Interleukin-1 alpha is released during transfection of keratinocytes. J Invest Dermatol 103: 580–582.

Konopka K, Davis BR, Larsen CE, Alford DR, Debs RJ, Duzgunes N (1990): Liposomes modulate human immunodeficiency virus infectivity. J. Gen. Virol. 71: 2899–2907.

Konopka K, Stamatatos L, Larsen CE, Davis BR, Duzgunes N (1991): Enhancement of immunodeficiency virus type 1 infection by cationic liposomes: the role of CD4, serum and liposome-cell interactions. J. Gen. Virol. 72: 2685–2696.

Kuo M, Chao M, Taylor J (1989): Initiation of replication of the human hepatitis delta virus genome from cloned DNA: role of delta atigen. J. Virol. 62: 1855–1861.

Lai M (1995): The molecular biology of hepatitis delta virus. Annu Rev Biochem 64: 259–286.

Leighton J, Brada Z, Estes LW, Justh G (1969): Secretory activity and oncogenicity of a cell line (MDCK) derived from canine kidney. Science 163: 472–473.

Lin M-F, DaVolio J, Garcia R (1993): Cationic liposome-mediated incorporation of prostatic acid phosphatase protein into human prostate carcinoma cells. Biochem Biophys Res Commun 192: 413–419.

MacGregor G, Caskey C (1989): Construction of plasmids that express E. coli beta-galactosidase in mammalian cells. Nucleic Acids Res 17: 2363–2365.

Mahato RY, Kawabata K, Takakura Y, Hashida M (1995a): In vivo disposition characteristics of plasmid DNA complexed with cationic liposomes. J Drug Target 3: 149–157.

Mahato RY, Kawabata K, Takakura Y, Hashida M (1995b): Physicochemical and pharmacokinetic characteristics of plasmid DNA/cationic liposome complexes. J Pharm Sci 84: 1267–1271.

Malone R (1989): mRNA transfection of cultured eucaryotic cells and embryos using cationic liposomes. Focus 11: 61.

Morris DH, Schalch DS, Monty-Miles B (1981): Identification of somatomedin-binding protein produced by a rat hepatoma cell line. FEBS Lett 127: 221–224.

Nakabayashi H, Taketa K, Miyano K, Yamane T, Sato J (1982): Growth of human hepatoma cell lines with differentiated functions in chemically defined medium. Cancer Res 42: 3858–3863.

Neumann J, Morency C, Russian J (1987): A novel rapid assay for chloramphenicol acetyltransferase gene expression. Biotechniques 5: 444.

Roner MR, Sutphin LA, Joklik WK (1990): Reovirus RNA is infectious. Virology 179: 845–852.

Ryu W-S, Netter HJ, Bayer M, Taylor J (1993): Ribonucleoprotein complexes of hepatitis delta virus. J Virol 67: 3281–3287.

Sanes JR, Rubenstein JL, Nicolas JF (1986): Use of recombinant retrovirus to study post-implantation cell lineage in mouse embryos. EMBO J 5: 3133–3142.

Sells MA, Li J, Chernoff J (1995): Delivery of protein into cells using polycationic liposomes. Biotechniques 19: 73–78.

Sternberg B, Sorgi FL, Huang L (1994): New structures in complex formation between DNA and cationic liposomes visualized by freeze-fracture electron microscopy. FEBS Lett 356: 361–366.

Tagawa M, Yokosuka O, Imazeki F, Ohto M, Omata M (1996): Gene expression and active virus replication in the liver after injection of duck hepatitis B virus DNA into the peripheral vein of ducklings. J Hepatol 24: 328–334.

Thorsell A, Blomqvist AG, Heilig M (1996): Cationic lipid-mediated delivery and expression of prepro-neuropeptide Y cDNA after intraventricular administration in rat: feasibility and limitations. Regulatory Peptides 61: 205–211.

Wu TT, Bichko VV, Ryu WS, Lemon SM, Taylor JM (1995): Hepatitis delta virus mutant: effect on RNA editing. J Virol 69: 7226–7231.

Xu Y, Szoka FC Jr. (1996): Mechanism of DNA release from cationic liposome/DNA complexes used in cell transfection. Biochemistry 35: 5616–5623.

APPENDIX: MATERIALS AND SUPPLIERS

Materials	Suppliers
[^{14}C]-Butyryl coenzyme A	Amersham
BSA	Gibco BRL
CAT enzyme standards	Gibco BRL
Chloramphenicol	Gibco BRL
Dimethylformamide	Sigma
DDAB	Gibco BRL
DMEM, 1×, liquid.	Gibco BRL
DOPE	Gibco BRL
DOTMA	Gibco BRL
Econofluor™	Gibco BRL
Fetal bovine serum	Gibco BRL
Formaldehyde	Sigma
General inorganic chemicals, NaCl, etc.	Sigma
Glutaraldehyde	Sigma
LipofectACE	Gibco BRL

APPENDIX (*Continued*)

Lipofectamine	Gibco BRL
Lipofectine	Gibco BRL
Multi-well plates	Corning-Costar
OPTI-MEM 1 reduced serum media	Gibco BRL
PBSA	Gibco BRL
Penicillin-streptomycin, 100×, liquid	Gibco BRL
Petri dishes	Corning-Costar
Potassium ferricyanide	Sigma
Potassium ferrocyanide	Sigma
Scintillation vials	Corning-Costar
Sterile tubes (Protocol 3.1c)	Corning-Costar
Trypsin	Gibco BRL
X-gal	Gibco BRL
Triton X-100	Sigma

12

Transfection of Mammalian Cells with Yeast Artificial Chromosomes

William M. Strauss

Department of Medicine, Division of Gerontology, Beth Israel Hospital, 330 Brookline Avenue, Boston, Massachusetts 02215

DNA Transfer to Cultured Cells, Edited by Katya Ravid and R. Ian Freshney.
ISBN 0-471-16572-7   © 1998 Wiley-Liss, Inc.

Portions of this chapter have been adapted and reprinted with permission from a chapter entitled "Transfection of Mammalian Cells via Lipofection" from the book entitled YAC Protocols (© Humana Press, 1995).

1. INTRODUCTION

The purpose of this chapter is to provide the investigator with the techniques necessary for transfecting yeast artificial chromosomes (YACs) into mammalian cells. This may be considered under four general headings: preparation of YAC DNA, gel purification of YAC DNA, introduction of YAC DNA into mammalian cells, and analysis of transfectants. The Introduction will give an overview of each of these topic areas; the reagents used and detailed protocols follow in subsequent sections.

1.1. Preparation of YAC DNA

Large amounts of purified YAC DNA are required for transfections. Attention must be paid to the growth and embedding conditions in order to recover enough material to use total yeast DNA for gel purification. In this section techniques for the growth and handling of large amounts of healthy yeast culture are discussed. A fresh inoculum of cells is prepared to start a large 24-h culture. After harvesting this large volume of culture, the cellular DNA is processed into intact chromosomal DNA. To maintain its structure the chromosomal DNA is embedded at high density in low-melting-point agarose. This section contains the most important and technically difficult step in this chapter.

1.1.1. Background

The use of bakers yeast, *Saccharomyces cerevisiae,* for the cloning of extremely large genomic intervals (exceeding 1 mb) was made possible with the development of YACs [Burke et al., 1987].

YACs are linear molecules containing all the control elements necessary for stable replication and segregation during the yeast life cycle. This cloning strategy was used to develop a technology for shuttling large genomic intervals back and forth between mammals and yeast.

Growth conditions of yeast must be optimized for the recovery of quantitative amounts of YAC DNA for transfection. For the preparation of DNA, there are two important qualities of a yeast culture: the number of cells and the growth phase of the growing yeast. The growth conditions must ensure that a sufficient quantity of yeast cells from the early stationary phase are available. Older cells, which have entered the late stationary phase, are not suitable, as they have undergone maturation of their cell wall. After maturation of the cell wall, it becomes difficult to generate spheroplasts from the cells and the resulting chromosomal DNA preparation is of inferior quality. The procedure as outlined for growth of the culture is designed to yield a large amount of relatively synchronized cells at the appropriate stage of maturity.

1.1.2. Critical parameters

These preparations are expensive and time consuming. To be practical, quantitative recovery of transfection-quality gel-purified DNA requires that batch-to-batch variation be eliminated. Every chromosomal DNA preparation should have identical quantities of total yeast chromosomes. Two characteristics will determine the quality of the embedded DNA: the spheroplast density and the condition of the spheroplasts after embedding. The growth conditions experienced by the yeast will largely determine both the amount of cells and their condition.

One way to ensure consistent concentrations of cells is to define the growth rate of a YAC-bearing strain. The culture conditions should be defined to ensure that the variation in cell recovery is eliminated. Much of this variation can be eliminated by utilizing the inoculations and growth conditions outlined above in the previous section, at least as a starting point.

The final spheroplast concentration must be calculated for each preparation. One must determine, experimentally, the number of yeast cells recovered, and then project what the final embedding volume (plug volume) should be. This is accomplished by determining the ratio of initial cell volume to final plug volume. This ensures that the loading capacity of the pulsed field gel electrophoresis (PFG) apparatus is not exceeded.

The condition of the yeast's cell wall determines one's ability to recover intact yeast chromosomes or YACs from the yeast strain. In older cultures, a significant fraction of the cells are refractory to digestion with zymolyase-100T and are inferior to younger cells for chromosomal DNA preparations. In chromosomal plug preparations made from cultures with a large fraction of old cells, many of the embedded cells will not be lysed or be incompletely lysed. Sometimes plugs made from older cultures are also unstable as they may be contaminated by DNA-degrading activities.

1.1.3. Anticipated results

A 2-l preparation of most yeast strains should yield 20 ml of packed cells. This will supply 80–120 μg of intact YAC DNA. In a given transfection, the typical experiment requires approximately 40 μg of a 500-kb YAC.

1.2. Gel Purification of YAC DNA

1.2.1. Overview

The separation of the yeast genomic DNA from YAC DNA is necessary for purification. Currently, only one method is available that ensures reasonable purity and physical integrity. This method is called pulsed field gel electrophoresis (PFG) [Schwartz and Cantor, 1984; Chu et al., 1986] and involves the fractionation of the yeast/YAC DNA through a low-melting-point agarose gel. The region of the gel containing the YAC is excised and then used for subsequent experiments.

The PFG technique allows only a small window of optimal resolution for the separation of DNA species. The desired size range of DNA molecules must first be determined prior to the commencement of the DNA isolation. The agarose plugs that contain the entire deproteinated yeast genome are loaded into a low-melting-point agarose gel in which a slot well has been cast. The plugs are laid end to end in the well to span the gel. The PFG is then run using the desired run parameters. After the run, the right and left edges of the gel are sliced off, stained with ethidium bromide, and photographed alongside a ruler using UV light. The unstained portion of the gel is then compared to the stained and photographed portion of the gel. The region of the gel containing the DNA of interest is then sliced from the unstained gel, and the gel slice is equilibrated with 20 mM Tris pH 7.6, 1 mM EDTA/spermine before being used for further experimentation.

1.2.2. Background information

Purification of high-molecular-weight DNA using PFG is a reasonable method for the significant enrichment of a particular set of DNA species. The PFG environment can favor the isolation of a single molecular species. The environment can also exclude a certain size class while concentrating the remaining molecular species into a focused band. PFG-purified DNA has been used as a source for YAC library construction, for fluorescence *in situ* hybridization (FISH) probes, and in transfection of mammalian cells [Strauss et al., 1992, 1993; Strauss and Jaenisch, 1992].

1.2.3. Critical parameters

The critical parameters can be grouped into two categories: DNA preparation (previous section) and PFG run conditions.

DNA Preparation

The success of the whole protocol is determined absolutely by the quality of the input DNA. If the DNA is degraded or if the density is too high there will be significant contamination and poor resolution. Time is well spent preparing the highest quality DNA possible

PFG Run Conditions

Applied electric field angle proves to be a critical determinant of resolution at high DNA loading concentrations. Field angles of 120° are quite sufficient for low concentrations of chromosomal DNA. However, at cell loading concentrations above 0.28 (see Sections 4.2 and 4.3), the field angle must be lowered. If one does not lower the field angle the resolution between bands will often be compromised and result in a smeared gel. The optimal applied electric field angle can vary a little, but should be between 104° and 110°. Routinely we work with field angles of 105°–107°. One additional benefit of a decreased electric field angle is that the length time for a PFG run can be reduced.

At high DNA loading concentrations the agarose concentration of the PFG gel has a marginal effect on overall resolution. However, it does prove to have a major effect on overall DNA mobility. For a given DNA concentration the variation of PFG gel-agarose concentration over a three-fold range can result in a 25% change in mobility, although the relative separation of each chromosomal band will not change accordingly. Furthermore, high PFG agarose gel concentrations can complicate the recovery of the YAC DNA for transfection, while very low agarose concentrations can produce a gel that is very fragile (see Sections 4.3, 4.4, and 4.5). Thus we routinely work with agarose concentrations in a range of from 0.7% to 1.0%. We have successfully worked with concentrations as low a 0.5%, and as high as 1.2%. For most purposes a concentration of 0.8% is satisfactory.

The PFG switching routine is very important to isolate effectively the particular classes of DNA molecules for further experimentation. PFG gels can be either run with fixed switching times or with ramped switching times. The main difference between these two types of routines for preparative purposes is that fixed switching routines tend to exclude certain molecular size ranges and focus others. Ramped routines tend to spread resolution over a greater size range with a loss in ideal resolution for a particular class of sizes. Ideally if one is trying to concentrate all molecules over a certain size range, and exclude all molecules that are smaller, then a fixed switching routine is perfect. If however, one is trying to isolate a particular class of molecules, for example chromosome V from *S. cerevisiae*, then a ramped routine is preferred as both large and small molecules can be spread out over a greater portion of the gel.

1.2.4. Anticipated results

This protocol can produce quantities of DNA in the microgram range per milliter of PFG gel slice. The degree of contamination by DNA from other size species depends upon the quality of the DNA and the amount of DNA loaded upon the gel. The more dilute the DNA the lower the contamination due to comigration of different molecular species.

The time required to prepare sample DNA is variable. The actual PFG run time may be as short as 12 h or as long as 40 h. Runs longer than this can be achieved if one is working with very large molecules (>5 mb) but the time required to isolate DNA for further experimentation can become prohibitive.

1.3. Introduction of YAC DNA into Mammalian Cells

1.3.1. Overview

Efficient transfection of mammalian cells using DNA conjugated to cationic lipids was reported using DOTMA (see Section 4.1) and the process was termed lipofection [Felgner et al., 1987] to distinguish this method from other transfection procedures. With conventionally sized DNA molecules, and a variety of currently available cationic lipids, a wide range of efficiencies have been reported. Differing cellular targets respond to a particular lipofection protocol with significantly divergent results; thus no single protocol can guarantee universally optimized success. The scientist must try a variety of lipids, in combination with different cells and DNA.

Classification of cells morphologically into two groups can assist in determining the best approach to transfection. Cells in culture either grow in suspension or adhere to a substrate. Cells that grow in suspension grow with a minimal surface exposed to the environment. Adherent cells can either spread out to expose a maximal surface area or they can round up to expose a minimal surface to the environment. This difference in cellular topology is a determinant in the choice of a lipofection protocol. The first protocol described below is designed to work with adherent cells that spread out on a substrate. The second protocol is for cells grown in suspension or those which are adherent but exposing the minimal surface area.

Ideally, the DNA concentration can be varied to optimize the lipid : DNA ratio in the transfection complex. With conventionally sized DNA molecules this is easily accomplished. However, with ultrahigh molecular weight DNA, in the form of YACs, it is much more difficult to vary the DNA concentration. Purified YACDNA was first introduced into mammalian cells [Strauss and Jaenisch, 1992; Strauss et al., 1993] by simply excising a portion of a low-melt PFG gel, treatment of the gel with agarase, mixing the resulting slurry with lipid, and applying the mixture to adherent cells. By this method the DNA concentration is limited to the loading capacity of the PFG system employed. Methods have been described [Lamb et al., 1993] for providing some limited concentration but these never concentrate the material more than a fivefold. These concentration procedures can impair the quality of the input DNA; consequently the investigator must weigh the effort of preparation against the possible results. The protocols described below assume that the DNA is not concentrated further after gel purification.

1.3.2. Background information

Transfection of gel-purified YAC DNA represents a flexible approach to functional testing of large cloned genomic intervals. Lipofection's flexibility is the result of three features: (i) Due to the wide array of commercially available cationic lipids many cell types from differing species can be transfected with success. For instance, although cells derived from human sources are known to be difficult to transfect via fusion protocols, these cells can be successfully transfected with DNA-lipid micelles. (ii) Lipofection operates optimally at low DNA concentrations, as is the situation encountered with YAC transfections. Due to the loading capacity of PFG gels, the quantities available after purification of YAC DNA range from 1—5 μg per 2-ml of gel slice. (iii) The technology required to establish Lipofection in a laboratory is very modest. Assuming that the expertise and facilities to work with ultrahigh molecular weight DNA are available, all that is required is access to a standard tissue culture facility.

1.3.3. Critical parameters

The most important parameters can be divided into those that relate to the DNA-lipid transfection complex, and those that relate to the cells. Transfection of intact YACs relies on the gentle handling of the DNA before and after liquefaction of the agarose. During the dialysis of the agarose-bound DNA, attention must be directed to the quality of the water used to make buffers; it should be purified and sterile, e.g., by deionization followed by double glass/quartz distillation (DDW).

Condensing agents must be used with ultrahigh molecular weight DNA to prevent shearing. In general, a polyvalent cation will coordinate DNA and compact it. The coordination occurs through the negatively charged phosphate backbone of DNA and a positively charged repeating unit on the condensing agent. The most commonly used condensing agents are the polyamines, spermidine and spermine. Spermidine has a coordination number of 3 and spermine 4. The use of mixtures of spermidine/spermine have been used by several investigators in the production of YAC libraries. Spermine has about a 10-fold greater capacity for condensation than does spermidine [Gosule and Schellman, 1978]. Thus in the protocols that we have developed, the use of spermidine has been omitted during dialysis. The concentration of spermine was reduced to a minimum, as, in our experience, no additional benefit was observed with concentrations of spermine greater than 0.5 M. The spermine should always be from a fresh source to ensure that it is not oxidized.

A second condensing agent, poly-L-lysine (PLL), with a very high coordination number (>1000), was used during the digestion of agarose. While the binding of spermine is reversible under the conditions of high salt or of electric field strengths of 5–10 V/cm, the binding of PLL is essentially irreversible. It is very important to use the PLL very sparingly, as overtitration will result in the precipitation of DNA into a large stringy mass. A small volume of PLL is pipetted onto the agarose slice, prior to heating to 65°C–68°C. *Never agitate the sample while melting the agarose, as this will shear the DNA.* After cooling the melted gel slice to 40°C, agarase is added. At this point very gentle tapping of the mix can be performed and is sufficient to disperse the enzyme; NEVER VORTEX. Similarly, when adding lipid, gentle tapping of the tube is sufficient to disperse the lipid. A polystyrene tube is recommended to avoid adhesion of the lipid and DNA to the side of the tube. Finally, always use wide-bore pipettes when transferring the condensed DNA or DNA-lipid complex.

There is much lore concerning the transfectability of mammalian cells. Adherent cells occasionally show toxicity with certain lipids, for instance DOTMA has a very steep toxicity curve due to the fact that the formulation may be difficult for cells to metabolize. This toxic effect is ameliorated by transfecting a confluent monolayer. However, confluent monolayers are not ideal for transfection, as transfections are best performed on cells that are growing rapidly. Hence it is desirable for adherent cells to be subconfluent. Optimally, the cells should have almost reached confluence (90–95%) just prior to addition of the transfection complex.

Suspension cells should not reach plateau prior to transfection, as in most plateau suspension cultures many of the cells may be dead or apoptosing. Suspension cultures should therefore be used in mid-log phase, using a larger volume to compensate for the lower cell numbers.

1.3.4. Anticipated results

Transfection efficiency is dependent on cell type. Even if the protocol is optimized, the range of efficiencies can vary over several orders of magnitude. Utilizing murine fibroblastic cell lines and the adherent-cell transfection protocol, transfection efficiency can range from 10^{-5} to 10^{-6} drug resistant clones. Using some embryonal carcinoma cell lines with the suspension cell protocol, similar results can be obtained. With ES cells the range varies from 10^{-6} to 10^{-7} drug-resistant clones.

From this population of drug-resistant clones, a portion will contain intact YACs and some clones will contain YAC fragments. With attention paid to condensing the YAC DNA prior to transfection, this ratio can be invariant as a function of YAC size. For YAC clones above 100 kb, 10% of drug-resistant cell clones should contain intact YACs after transfection.

1.4. Analysis of Transfectants

1.4.1. Overview

To determine whether a transfected cell line contains a YAC requires screening many cell lines. This determination can present a significant technical problem. Several approaches exist: (i) drug selection, (ii) restriction mapping, and (iii) *in situ* hybridization. A complete discussion of these topics is outside the context of this chapter. Some general comments are presented, as well as a useful protocol.

Most conventional methods for isolating genomic DNA utilizes a proteinase K step, followed by phenol/CHCl$_3$ extraction to remove protein. The DNA produced by this type of method is very pure and stable in storage. Unfortunately the organic extractions require multiple pipetting steps and transfer of aqueous DNA solution to several different tubes. When dealing with many samples, this represents a prohibitive amount of work.

The protocol described here does not involve any organic extractions or tube changes [Laird et al., 1991] as the cells are lysed *in situ* in the growth vessel, a 24-well dish. This protocol enables a single investigator to produce restriction-enzyme-digestible genomic DNA from hundreds of cell lines with minimum effort. In order to determine whether transfected cell lines contain YAC DNA, genomic DNA must be isolated from many cell lines. The following protocol was designed to facilitate the isolation of small amounts of DNA for initial screening of hundreds of cell lines. The cell samples can be lysed *in situ* and DNA of sufficient quality for restriction digestion and standard electrophoresis can be isolated from individual wells of a 24-well tray.

1.4.2. Background information: Drug selection of transfectants

In order to differentiate the cells that have taken up YAC DNA from those that have not, drug selection must be used. The YAC can contain a *cis*-encoded drug resistance gene or a marker can be co-transfected [Strauss and Jaenisch, 1992; Strauss et al., 1993; Choi et al., 1993]. Most YACs are from libraries where no drug marker was inserted into the vector. Thus most YACs must be retrofitted with a resistance marker. The most important consideration is to use the marker system that will suit the experimental need best. A few general points are worth considering. If two different markers are used, one on each arm, then after transfection each arm of the YAC can be selected for. Only those clones with both arms of the YAC would therefore be further characterized. Some marker cassettes also contain rare restriction sites, which can be very useful in subsequent characterization. Finally, the location of the drug marker can be chosen with great precision. Different locations provide very different advantages. The investigator should carefully consider where to place the selectable marker(s).

Due to the nature of the DNA-lipid micelles formed during lipofection, co-transfections are possible. To achieve co-transfection it is necessary to mix a drug marker cassette in a limiting molar quantity with YAC DNA. The DNA-lipid micelle is then formed of two DNA species. The transfection complex is mixed with the target cell. Application of the drug to the media would then proceed as usual. As the marker DNA is present in limiting quantity compared to YAC DNA, the chances for both DNA species to coexist in the same cell line is improved. Despite the limiting amount of selectable marker DNA in the transfection complex, most of the drug-resistant colonies will not contain YAC DNA.

On the basis on the preceding discussion, the central disadvantage of the co-transfection approach is that the selection step is reduced to a screening step for YAC DNA. Many clones that are geneticin-resistant (G418^R) turn out not to contain any YAC-derived material. By using this screening step one must screen 10^2–10^3 clones instead of 10^7–10^8 cells in the whole culture. Certainly this is labor-saving but it represents ten times more work than using a colinear selectable marker.

The advantage of the co-transfection approach is that the YAC does not have to be modified. For some experiments this is a major advantage. For instance YACs can be quite unstable. If this situation occurs, then modification by homologous recombination may be prohibited. By co-transfection one can successfully transfect the gel-purified material despite the inability to modify the clone.

The central issue in transfections of YACs is the integrity of DNA after manipulation. DNA integrity can be impaired in the yeast host, during gel purification, or during the transfection itself. Given the very large size of YAC clones, the determination of YAC integrity represents a formidable problem.

There are three approaches to the determination of YAC integrity. The first approach utilizes a single probe that hybridizes to YAC DNA at many locations. One such probe would be a LINE sequence probe. This probe can be used in combination with a restriction enzyme that recognizes a 6-base sequence, to generate a YAC fingerprint. This fingerprint is very useful, especially after transfection. The fingerprint of the YAC before transfection is compared to the transfected YAC fingerprint. If there is significant similarity between the two fingerprints then there is a good chance that the YAC is intact.

The second approach relies on the availability of many unique sequence probes. These unique probes are derived specifically from different regions of the YAC. Hybridization of these probes to Southern blots from transfected cell lines indicates the presence (or absence) of particular regions of the YAC. This hybridization can be accomplished individually or in pools. A pooling strategy can significantly reduce the amount of work, and the resulting data will look much like the fingerprint generated with a repeat probe.

Both of the two preceding approaches can yield important information using infrequently or frequently cutting restriction enzymes. The choice of enzyme is largely dictated by the method of DNA preparation. If the high throughput DNA isolation protocol (see Section 3.4) is used, then a frequently cutting enzyme must be used with a standard electrophoretic environment.

If the YAC DNA can be differentiated from host DNA by restriction fragment length polymorphisms (RFLPs), then probes from the YAC vector or from the cloned insert can be used for structural analysis. For instance, if a human YAC clone is transfected into a murine cell line, then either of the two foregoing approaches to restriction mapping could be used.

On the other hand, if the degree of polymorphism between the YAC and host genome is not great, only the second approach may be useful. One example is the use of *Mus spretus* YAC clones to transfect *M. musculus* cell lines. In this case there is little repeat sequence divergence, so a repeat sequence fingerprint cannot be generated. There is sufficient sequence divergence, however, to differentiate a *spretus* YAC from a *musculus* genome with unique sequence probes and informative restriction enzymes. The appropriate choice of enzyme will differentiate the YAC sequences from the host sequences.

In either case, the vector probe will allow the investigator to determine the copy number of each arm in the transfected cell line. For instance, no Bam HI sites exist in the pYAC4 vector; thus, using this enzyme and a probe from one arm will show a single hybridizing band, if the YAC is present in single copy number. The arm-specific probes can be easily generated by the digestion of pBR322 with PvuII and Bam HI. The two fragments generated will each correspond to one arm.

A variant of restriction mapping for the purpose of determining the integrity of transfected YAC DNA depends upon engineering the YAC before transfection. If rare cutting sites are incorporated on either side of the YAC vector, then an even stronger demonstration of integrity after transfection can be made. If these rare cutting sites are not found in the cloned insert, after restriction digestion, fractionation, and hybridization, a unique restriction fragment the approximate size of the original YAC will be revealed only if the transfectant contains an intact YAC. Rearranged transfected YACs should exhibit a band of different mobility.

The third approach to the determination of transfected YAC integrity requires fluorescence *in situ* hybridization (FISH) of mitotic and interphase chromosomes of transfected cell lines to screen transfected cells rapidly for YAC DNA. One can determine which YAC fragments are present in the cell line, which fragments are colinear, and where in the genome the YAC transgene is located. As RFLPs are not needed, no restriction enzymes are required. This is possible because FISH technology operates by the hybridization of fluorescently tagged DNA probes to chromosomal DNA. The hybridization target can either be mitotic or interphase chromosomes. YAC-specific probes can be rapidly screened by this procedure with a point-to-point resolution of <10 kb when using interphase chromosomes.

1.4.3. Critical parameters

Proteinase K is a very robust and stable enzyme. This protocol uses the minimal amount that still allows for isolation of high-quality DNA. If the investigator exceeds the amounts specified, proteinase K activity will still be found in the processed DNA. This carryover of contaminating proteinase K activity will prevent the digestion of the DNA with restriction enzymes and the DNA will be useless for analysis by Southern blotting and hybridization.

During the overnight digestion of the DNA with proteinase K, the samples must be mixed. The consistent gentle agitation of the tissue culture plate ensures that the cells are completely dispersed and digested. If the plate is not rocked, there will be incomplete processing and the DNA will fail to digest well.

The 24-well plate can contain varying numbers of cells, depending on the nature of the cell line. It is important to ensure that each well has grown up to confluence, or near confluence. It is also important that there be little well-to-well variation; if the number of cells is consistent then the recovery of DNA will be similar from well to well.

1.4.4. Anticipated results

After the cells are grown up in the 24-well plates, it is expected that high-grade DNA will be available in less than 2 days. The difference between working with one 24-well plate and 10 plates is minimal. Furthermore, the yield of DNA does not vary with the scaling of the experiment. In general, one can expect 100–200 μg of DNA to be recovered from each well, enough for about 10 restriction digests or 10 lanes on a gel.

2. PREPARATION OF REAGENTS

2.1. Preparation of Large-Volume Yeast Cultures and High-Density Chromosomal DNA

2.1.1. Digestion buffer

EDTA .... 250 mM
Tris HCl pH7.6 .... 20 mM
n-Lauryl sarcosine 2%
Proteinase K to .... 0.5 mg/ml

Add the proteinase K just before use.

2.1.2. Phenylmethylsulfonylfluoride (PMSF) 100× stock

PMSF, 0.1 M in 100% isopropanol, made up fresh prior to use.

2.1.3. Storage buffer

EDTA .... 0.1 M
Tris HCl pH 8.0 .... 0.02 M

2.1.4. Agarose

Sea Plaque-GTG agarose, 5%, in storage buffer

2.1.5. Zymolyase 100T

Powder

2.1.6. Sorbitol

Sorbitol, 1 M, in 20 mM Tris HCl, pH 7.0.

2.1.7. Dropout mix powder

(Mixture lacking uracil and tryptophan for selection of conventional YACs)

Adenine .... 2 g
Alanine .... 2 g
p-Aminobenzoic acid .... 0.2 g
Arginine .... 2 g
Asparagine .... 2 g
Aspartic acid .... 2 g
Cysteine .... 2 g
Glutamic acid .... 2 g
Glutamine .... 2 g
Glycine .... 2 g
Histidine .... 2 g
Inositol .... 0.1 g
Isoleucine .... 2 g
Leucine .... 4 g
Lysine .... 2 g
Methionine .... 2 g
Phenylalanine .... 2 g
Proline .... 2 g
Serine .... 2 g
Threonine .... 2 g
Tyrosine .... 2 g
Valine .... 2 g

Mix together well and store at room temperature.

2.1.8. Rich dropout medium

Plates
- Glucose .... 40% in UPW
- Flask 1 (500 ml)

Dropout mix .... 2 g
Yeast nitrogen base, without amino acids and
ammonium sulfate .... 1.45 g
Ammonium sulfate .... 5 g
UPW .... 500 ml

- Flask 2
  Suspend 20 g agarose (see Section 2.1.4) in 450 ml of UPW.
- Autoclave each flask separately and then mix:

Flask 1 .... 500 ml
Flask 2 .... 450 ml
Glucose, 40% .... 50 ml

Pour plates and allow to set for 24 h prior to use.

Liquid medium

Liquid medium is prepared in a similar fashion except that the agar is omitted.

2.1.9. β-Mercaptoethanol

(commercially available analytical grade)

2.2. Gel Purification

2.2.1. Tris-borate-EDTA (TBE) 10×

Tris base .... 242 g
Boric acid .... 123 g
EDTA-Na$_4$ .... 5.2 g
DDW .... 4 liters

Dilute 1:20 prior to use in a PFG apparatus, 1:10 for other electrophoretic applications.

2.2.2. Low-melt agarose

Sea Plaque GTG agarose

2.2.3. Ethidium bromide

Ethidium bromide, 10 mg/ml stock solution in distilled H$_2$O

2.2.4. Dialysis buffer (TSE)

Tris HCl, pH 7.6 .... 20 mM
EDTA .... 1 mM
Spermine .... 0.1 M

2.2.5. Tris-borate-EDTA (TBE), 10×

Tris base .... 242 g
Boric acid .... 123 g

EDTA-Na$_4$ .... 5.2 g
DDW .... 4 l

2.2.6. Pulsed field gel apparatus

2.3. Lipofection

2.3.1. Cell culture media

Dulbecco's modification of Eagle's medium (DMEM), 10×
OptiMEM
Evans-Kaufman (EK) medium:

Fetal bovine serum .... 75 ml
Nonessential amino acids .... 5 ml
Penicillin/streptomycin .... 5 ml
β-Mercaptoethanol .... 4 ml
High-glucose DMEM to .... 500 ml

2.3.2. Commercially available lipids (see Section 4.1)

Lipofectin (DOTMA and DOPE)
Lipofectamine (DOSPA and DOPE)
DOTAP
DOGS

2.3.3. Condensing agents

Spermine
Poly-L-lysine, low molecular weight (Sigma P-8954)

2.3.4. β-Agarase

For transfections it is important to use a very high quality agarase. Two preparations, NEB and Gelase, are recommended.

2.4. Genomic DNA Isolation

2.4.1. Lysis buffer

Tris HCl, pH 8.5 .... 0.1 M
EDTA .... 5 mM
SDS 0.2%

On the day of the procedure add proteinase K stock to produce a concentration of 100 μg/ml.

2.4.2. Proteinase K stock solution (1000×)

Tris HCl, pH 8.0. .... 20 mM
EDTA .... 1 mM
Proteinase K .... 100 mg/ml

3. METHODS

3.1. Preparation of Large-Volume Yeast Cultures and High-Density Chromosomal DNA

Protocol 3.1

Reagents and Materials
- Distilled, deionized water or equivalent (UPW)
- Sorbitol, 1 M
- Rich dropout medium
- Roller drum
- Shaking incubator
- Beckman J6 centrifuge or equivalent
- Beckman 1-liter canister bottles
- Conical centrifuge tubes, 50 ml, plastic
- β-Mercaptoethanol
- Zymolase 100-T
- Conical centrifuge tubes, 3 × 50 ml
- SeaPlaque GTG low-melt agarose, 5% in 0.1 M EDTA
- Digestion buffer
- PMSF, 1 mM

Protocol
(a) Inoculate 50 ml of rich dropout medium with a 1:100 dilution of a stationary culture.

(b) Grow the culture overnight at 30°C in a roller drum to reach saturation density.

(c) Inoculate 2 l of the same dropout medium with the fresh overnight culture at an inoculation of 1:100.

(d) Shake cultures at 260 rpm at 30°C. (Grow up for a duration not to exceed 24 h. It is important not to overgrow the culture).

(e) Transfer the cells and broth to Beckman 1-liter canister bottles and spin in J6 centrifuge at 800–1000 g for 10 min, to pellet the cells.

(f) Resuspend the cells in DDW and repeat the centrifugation.

(g) Resuspend the cells in 1 M sorbitol and repeat the centrifugation.

(h) Resuspend the cells in 10 ml of 1 M sorbitol and transfer to a 50-ml plastic conical centrifuge tube. Rinse out the 1-liter canister to recover all of the cells and add to the 50-ml tube.

(i) Spin down the cells in a J6 centrifuge at 800–1000 g for 5 min and gently remove the clear supernatant. Out of a total volume of 30 ml, the sorbitol will contribute 10 ml, and there should be approximately 20 ml of cells (see Section 4.2).

(j) Add 17 ml of 1 M sorbitol and 1 ml of β-mercaptoethanol to bring final volume to 48 ml.

(k) Add 60 mg of zymolase 100-T and mix.

(l) Split the reaction between three 50-ml conical tubes (16 ml per tube) and incubate at 37°C for 45–70 min until free spheroplasts are seen (check microscopically, see Section 4.11).

(m) Add 5 ml of 5% SeaPlaque GTG low-melt agarose in 0.1 M EDTA (equilibrated to 54°C) and immediately cast.

(n) Eject the plugs into two fresh 50-ml plastic conical centrifuge tubes, using an air line.

(o) Add lysis buffer (with fresh proteinase K) and digest overnight at 50–55°C with *very gentle* agitation.

(p) Treat with 1 mM PMSF by simply adding a 1 : 100 volume of stock solution to the tube and letting it sit for 60 min at room temperature.

(q) Decant the inactivated lysis buffer and add storage buffer.

3.2. Gel Purification of YAC DNA

Protocol 3.2

Reagents and Materials

Aseptically prepared

- TBE, 0.5×: dilute from 10× TBE (see Section 2.2.5)
- Agarose in 0.5× TBE, 1.2%
- DNA plugs
- DNA size markers
- Ethidium bromide, 0.1 mg/ml in 0.5× TBE
  Note: Carcinogenic; handle with care
- TSE dialysis buffer (see Section 2.2.4)
- Conical centrifuge tubes, 50 ml
- Polystyrene tubes, 15 ml, round bottomed

Non Sterile

- Gel casting plate

Non Sterile Equiptment

- Pulsed field gel electrophoresis (PFG) apparatus
- UV transilluminator (short- or medium-wavelength UV is suitable)

Protocol

(a) Prepare source DNA in agarose plugs, and dialyze in 0.5× TBE

(b) Pour a 0.5–1.2% low-melt agarose gel in 0.5× TBE gel using a glass casting plate with Velcro strips. Use a single-slot trough with small wells at either end for lane markers (see Section 4.5).

(c) Equilibrate the DNA plugs against running buffer (0.5× TBE) and load them into the slot trough lengthwise. Make sure that the plugs are as close to each other as possible without placing them under significant compression. The plugs can be sealed in the trough with a little warm agarose (use the same mixture as you used to pour the gel).

(d) Into the outer lanes, load size markers to ensure accurate sizing of the DNA fragments. (If one is fractionating whole yeast chromosomes the step is unnecessary).

(e) Place the whole glass plate (with the gel firmly attached to it) into a prechilled PFG apparatus. Let the gel equilibrate in the chamber for 10–15 min, then commence the PFG run.

(f) After the PFG run is complete, slice off the left and right edges of the gel containing the size markers and a small portion of the sample DNA. Carefully slide these samples into a staining bath containing 0.5× TBE with 0.1 mg/ml of ethidium bromide. Let the gel slice slices stain for 30–60 min, and destain with 0.5 TBE alone for 30–60 min (see Sections 4.6 and 4.7).

(g) Place these slices side by side next to a ruler atop a UV transilluminator and photograph. It is important to examine both sides of the gel as PFG gels often run with a little deformation from edge to edge.

(h) Going back to the remaining unstained gel, place two rulers on either side of the gel. Line up a third ruler across the gel using the photograph as a reference. Use a clean, sterile scalpel to slice a strip of the desired section of the gel.

(i) Transfer this slice to a 50-ml conical centrifuge tube and dialyze against TSE. At least three changes are required over a course of 12 h. The DNA is usually stable for several days to a week, but it should be used as soon as possible.

(j) In order to access the gel-bound DNA, subdivide the gel slice into 0.5-cm to 2.0-cm sections and transfer into capped and sterile 15-ml round-bottomed polystyrene tubes.

3.3. Lipofection Procedures

3.3.1. Lipofection of adherent cells

Protocol 3.3.1

Reagents and Materials
Aseptically prepared
- PLL: poly-L-lysine (see Section 2.3.3)
- β-Agarase
- Cationic lipid (e.g., DOTAP)
- OptiMEM
- Petri dishes, 100 mm

Protocol
(a) A few days (the time depending upon growth rate of cells) prior to the transfection, plate out cells onto 100-mm dishes. The plating should ensure that on the day of the transfection the cells will be 95% confluent (see Section 4.8).

(b) On the day of transfection add PLL to the gel slice to a final concentration of 4 μg/ml.

(c) Warm the tube to 65–68°C until the gel slice is melted, and equilibrate at 40°C for 5 min.

(d) Add 10 units of β-agarase per dish, gently mix, and allow to digest for 60–120 min at 40°C.

(e) Add the appropriate amount of cationic lipid (for DOTAP 5–30 μg) and allow the mixture to complex at room temperature for 30 min.

(f) Equilibrate the transfection complex sample with sterile DMEM prepared by diluting a 10× DMEM solution 10 fold in sterile distilled water. For some cells 1 ml of OptiMEM can be added at this stage. The final volume should be between 2 and 5 ml.

(g) Wash the monolayer of cells free of standard medium with Opti-MEM and then apply the transfection complex (see Section 4.9).

(h) Stop the transfection at 4 h, or continue overnight, depending on the viability of the cells. Stop the transfection by changing to fresh medium. Culture for 24–48 h at 37°C and apply selection.

3.3.2. Lipofection of cells in suspension

Protocol 3.3.2

Some cells grow in suspension, and some cells that grow attached are better transfected in suspension. In particular this protocol was developed for use with murine embryonic stem cells (ES cells).

Reagents and Materials
Aseptically prepared
• Frozen vial of ES cells
• Feeder layer, e.g., STO cells, 2×10^4 cells/ml in 25-cm^2 flask
• Irradiated feeder cells in 10-cm tissue culture dish
• Evans-Kaufman (EK) medium
• Polystyrene tubes, 15 ml
• Culture flasks, 75 cm^2

Protocol
(a) Thaw a vial of ES cells, seed onto a pre-established feeder layer in a 25-cm^2 flask in EK medium, and expand for 3 days.

(b) Split the flask 1 : 6 into three, 75-cm^2 flasks and culture for 3 days on feeder layers. Depending on the size of the experiment these three flasks can again be split into ten 175-cm^2 flasks (approximately 1 : 8 split by surface area). It is important to use the cells on the 3rd day after seeding.

(c) On day 3, trypsinize the cells and remove the feeder cells from the culture by preplating on fresh plastic plates for 30 min at 37°C. Collect the nonadherent cells, wash, and count.

(d) Add 5×10^6 to 1×10^7 cells to each 15-ml polystyrene tube containing the prepared transfection complex, loosely recap, and place in a CO_2 incubator at 37°C for 4 h with occasional (once per hour) gentle mixing to resuspend the settled cells.

(e) Subsequently plate the cells from each tube into a single 10-cm tissue culture dish over a monolayer of irradiated feeder cells in EK medium (see Section 2.3.1).

(f) After 24–48 h, change the medium and place the cells under selection. This is called day 0.

(g) By day 2 there should be massive cell death, and few colonies should remain. By day 6, resistant colonies should be discernible from background and colonies can be picked on day 8 to day 10 (see Section 4.10).

3.4. High Throughput Genomic DNA Isolation

Protocol 3.4

Reagents and Materials
- Growing culture of transfected cells in 24-well plate
- Lysis buffer (see Section 2.4.1)
- Storage buffer
- Isopropanol
- Rocking or rotating rack

Protocol

(a) Pipette 500 μl of lysis buffer into each well of a 24-well plate containing growing cells.

(b) Incubate overnight at 55°C on a gently rocking or rotating rack.

(c) Add 500 μl of isopropanol.

(d) Rock the samples back and forth to mix. After the samples are well mixed the DNA should precipitate and form a lacey sediment.

(e) With a clean pipette tip pick the sediment out of the well and transfer it to a fresh tube. The samples can be washed (optional) by centrifugation with 95% ethanol.

(f) When all tubes are complete add 100 μl of storage buffer to the samples and allow to resolubilize. Use 10 μl for each genomic digest destined for Southern blot.

4. NOTES

4.1. Lipids

DOTMA: *N*-[1-(2,3-diioleyloxy)propyl]-*N,N,N*-trimethylammonium chloride

DOSPA: *N*-[2-({2,5-bis(3-aminopropyl) amino]-1-oxypentyl}amino)ethyl]-*N,N*-dimethyl-2,3-bis(9-octadecenyloxy)-1-propanaminium trifluoroacetate)

DOTAP: *N*-[1-(2,3-dioleoyloxy)propyl]-*N,N,N*-
        trimethylammoniummethylsulfate
DOGS: dioctadecylamidoglycyl spermidine
DOPE: dioleoyl phosphatidylethanolamine

4.2. Agarose Plugs

All the proportions of sorbitol and agarose are calculated to achieve a certain concentration of cells, expressed as a ratio of initial cell volume to the final volume of agarose, cells, and sorbitol. The desired ratio of cell volume to final plug volume is between 0.32 and 0.35. For example, if one has 20 ml of cells and wants a final ratio of 0.32 then the final amount of agarose embedded cells would be 63 ml ($20/63 = 0.32$).

4.3. DNA Loading

Do not overload the gels, since the DNA density in the plugs makes a significant difference in final resolution on PFG. Routine concentrations of 6×10^8 to 1×10^9 yeast cells/ml are never a problem, very high concentrations of yeast can be used with care.

4.4. Casting Gels

If it takes a long time for the low-percentage agarose gels to solidify, cast the gel in a 4°C cold room.

4.5. Supporting the Gel During Electrophoresis

The gel sets into the Velcro and thus the casting plate grips and supports the delicate gel during handling and electrophoresis. This item is easily homemade using Velcro strips (only the teeth portion) silicone glue and a standard 20×20 glass plate.

4.6. Supporting the Gel During Staining

It facilitates further transfer of these staining samples if the strips are placed on top of a plastic or glass sheet prior to and during staining.

4.7. Destaining

Destaining stained gels in buffer improves the signal-to-noise ratio by reducing the backround.

4.8. Cell Density at Transfection

It has been observed that there is less toxicity with confluent monolayers (see also Section 1.3.3).

4.9. Manipulation of DNA

It is important to transfer the transfection complex as gently as possible to ensure that the DNA is not sheared. The use of wide-bore pipettes is recommended.

4.10. Transfection of Nonadherent Cells

As some cells grow in a nonadherent manner, different strategies must be employed for selection. Obviously if one is working with a cell line that grows in suspension the latter half of this protocol is inappropriate. After the transfection step (Protocol 3.3.2d) the cells can be selected in bulk, after a significant period of drug selection, and cloned in microtitration plate wells. This approach selects for faster growing clones which presents a disadvantage, as clones that grow faster will tend to predominate and bias recovery of other clones. An alternative is cloning or simple fractionation of the transfected cells prior to the commencement of selection.

4.11. Visualization of Spheroplasts

Microscopic determination of spheroplast formation is best achieved in the following manner: Prepare 1% SDS and 1 M sorbitol solutions. On either end of a microscope slide place either a 10 μl drop of SDS or a 10 μl drop of the sorbitol. To each drop add 2.5 μl of the treated yeast culture. Mix and cover the drops with a coverslip, keeping them separated from one another. Examine the cells under a phase microscope at 200× magnification. The sorbitol-treated cells should look unchanged from untreated cells, the SDS-treated cells should appear blown apart, pale, and ghostlike if spheroplasts have been formed successfully.

REFERENCES

Burke DT, Carle GF, Olson MV (1987): Cloning of large segments of exogenous DNA into yeast using artificial-chromosome vectors. Science 236: 806–812.
Choi TK, Hollenbach PW, Pearson BE, Ueda RM, Weddell GN, Kurahara CG, Woodhouse CS, Kay RM, Loring JF, et al. (1993): Transgenic mice containing a human heavy chain immunoglobulin gene fragment cloned in a yeast artificial chromosome. Nature Genet 4: 117–123.
Chu G, Vollrath D, Davis R (1986): Separation of large DNA molecules by contour-clamped homogeneous electric fields. Science 234: 1582–1585.
Felgner PL, Gadek TR, Holm M, Roman R, Chan HW, Wenz M, Northrop JP, Ringold GM, Danielson M, et al. (1987): Lipofection: a highly efficient, lipid-mediated DNA-transfection procedure. Proc Natl Acad Sci USA 84: 7413–7417.
Gosule LC, Schellman JA (1978): DNA condensation with polyamines. J Mol Biol 121: 311–326.

Laird PW, Zijderfeld A, Linders K, Rudnicki MA, Jaenisch R, Berns A, et al. (1991): Simplified mammalian DNA isolation procedure. Nucleic Acids Res 19: 4293.

Lamb BT, Sisodia SS, Lawler AM, Slunt HH, Kitt CA, Kearns WG, Pearson PL, Price DL, Gearhardt JD, et al. (1993): The introduction and expression of the 400 kb human genomic sequence of the amyloid precursor protein gene in transgenic mice. Nature Genet 5: 22–30.

Schwartz DC, Cantor CR (1984): Separation of yeast chromosome-sized DNAs by pulsed field gradient gel electrophoresis. Cell 37: 67–75.

Strauss WM, Jaenisch R (1992): Molecular complementation of a collagen mutation in mammalian cells using yeast artificial chromosomes. EMBO J 11: 417–422.

Strauss WM, Dausman J, Beard C, Johnson C, Lawrence B, Jaenisch R, et al. (1993): Germ line transmission of a yeast artificial chromosome spanning the murine alpha1(I) collagen locus. Science 259: 1904–1907.

Strauss WM, Jaenisch E, Jaenisch R (1992): A strategy for rapid production and screening of yeast artificial chromosome libraries. Mamm Genome 2: 150–157.

APPENDIX: MATERIALS AND SUPPLIERS

Material	Supplier
Agarase	New England Biolabs
Agarase, gelase	New England Biolabs
Ammonium sulfate	Difco
DOGS	Promega
DOTAP	Boehringer Mannheim
Dulbecco's modification of Eagle's medium (DMEM), 10×	Gibco-BRL
Evans-Kaufman (EK) medium	Gibco-BRL
Fetal bovine serum	Gibco-BRL
High-glucose DMEM	Gibco-BRL
Lipofectamine (DOSPA, DOPE)	Gibco-BRL
Lipofectin (DOTMA, DOPE)	Gibco-BRL
Nonessential amino acids	Gibco-BRL
OptiMEM	Gibco-BRL
Penicillin 10,000 U/ml; streptomycin, 10 mg/ml	Gibco-BRL
Phenylmethylsulfonylfluoride (PMSF)	Sigma
Poly-L-lysine, low molecular weight	Sigma
Polystyrene tubes, 15 ml, round-bottomed	Falcon
Proteinase K	Sigma
Pulsed field gel apparatus (CHEF DRIII)	Biorad
Sea Plaque GTG agarose	FMC
Sorbitol	Sigma
Spermine	Fluka
Yeast nitrogen base without amino acids and ammonium sulfate	Difco
Zymolase 100-T	Seigaku

13

Mapping Human Senescence Genes Using Microcell-Mediated Chromosome Transfer

Robert F. Newbold and Andrew P. Cuthbert

Human Cancer Genetics Unit, Department of Biology and Biochemistry, Brunel University, Uxbridge UB8 3PH, United Kingdom

DNA Transfer to Cultured Cells, Edited by Katya Ravid and R. Ian Freshney.
ISBN 0-471-16572-7 © 1998 Wiley-Liss, Inc.

Editors' Note: This chapter was originally contributed to a previous volume in the series, "Culture of Immortalized Cells," Freshney, R. I. & Freshney, M. G. (1996) Wiley, New York. We apologize for reproducing it here but, unfortunately, our alternative choice of author for a chapter on whole chromosome transfer was unable to complete the chapter for personal reasons, and, as we felt that this volume would be incomplete without it, we asked Prof. Newbold and Dr. Cuthbert if we could reuse their previous chapter; they agreed and it is reproduced here with slight modifications.

I. BACKGROUND

All adult mammalian cells derived from tissues capable of proliferation *in vivo* display a limited proliferative capacity *in vitro*. The resulting loss of division potential is usually termed "replicative senescence" [Goldstein, 1990]. This, apparently preprogrammed, limitation to cellular growth presents a formidable barrier to clonal evolution and malignant transformation in culture and may constitute an important tumor suppressor mechanism [Newbold et al., 1982; Newbold and Overell, 1983; Newbold, 1985; Trott et al., 1995a]. How mammalian cells control their mitotic potential so precisely is not known, although a very attractive mechanism to account for cell senescence and immortalization has recently been proposed based on telomere shortening and telomerase reactivation, respectively [Counter et al., 1992, 1994; Grieder, 1990]. A better understanding of the cell and molecular biology of these two alternative cellular proliferation phenotypes should contribute to our understanding of fundamental cancer mechanisms. In addition, improvements in knowledge in this area may also suggest means by which specialized mammalian cell types, particularly of human origin, might be efficiently immortalized, while at the same time maintaining their differentiated functions. Such an advance would, in turn, be likely to facilitate studies of the biology of mammalian cell differentiation and would provide cellular systems for the commercial production of differentiated cell proteins (e.g., growth factors and hormones) for medical and other uses. It is to the latter interest groups that the contents of this chapter are primarily directed.

Microcell-mediated monochromosome transfer is a powerful new technique in modern somatic cell genetic analysis [Saxon and Stanbridge, 1987; Cuthbert et al., 1995]. The method permits a single chromosome (for example, a normal human chromosome) to be introduced into almost any homologous or heterologous mammalian cell background, and maintained therein (by selection) as an intact functional and structural entity. Using this technique, individual chromosomes can be screened for the presence of active genes specifying a particular cellular phenotype (e.g., those controlling the replicative senescence programme) on the basis of genetic complementation.

Heterospecific human:rodent microcell hybrids offer a major technical advantage over homospecific combinations in genetic complementation studies, in the sense that it is a relatively straightforward task to define the structural integrity of the foreign chromosome by PCR analysis of sequence-tagged-site (STS) genetic markers and chromosome painting. This facilitates deletion mapping of the introduced chromosome in rare candidate segregants that do not express the phenotype of interest. Such heterospecific transfers, involving the introduction of normal human chromosomes singly into newly immortalized rodent cells, have led recently to the identification of genes which may be important in controlling replicative senescence and their assignment to distinct human subchromosomal regions [e.g., Hensler et al., 1994; Cuthbert et al., 1995]. However, the differential susceptibility of cultured rodent and human cells to spontaneous or carcinogen-induced immortalization [Trott et al., 1995a; Russo et al., 1998] suggest that control over replicative potential is substantially more robust in humans and therefore, functional genetic complementation studies should be conducted using both heterospecific and homospecific hybrids. The following account describes in detail the procedures used in such analyses and summarizes additional technical developments (Sections 7.3 and 8) that are currently being employed to isolate the genes responsible for senescence induction.

2. DERIVATION OF NEWLY IMMORTALIZED SYRIAN HAMSTER CELL LINES FOR USE AS RECIPIENTS IN MICROCELL TRANSFER EXPERIMENTS

2.1. Initiation of Primary Cultures from Syrian Hamster Dermis (SHD)

In general, normal rodent cells in culture readily generate immortal variants, unlike their human counterparts, which are completely resistant to spontaneous immortalization and can only be induced to generate immortal variants (by carcinogens or DNA tumor virus early genes) with great difficulty [Bai et al., 1993]. Rodent cells from different species, however, differ greatly in their propensity to immortalize. For example, Syrian hamster cells (e.g., fibroblasts) are markedly more resistant to spontaneous immortalization than counterparts obtained from rat or mouse [Newbold et al., 1982; Trott et al., 1995a]. Moreover, in contrast to human cells, immortal variants of Syrian hamster cells can be readily induced against a zero spontaneous background by treating cultures with chemical carcinogens or ionizing radiation x-rays, (γ-rays, or fast neutrons). In addition to providing a suitable system for quantifying the immortalizing efficiencies of carcinogens, an immortalization assay based on Syrian hamster fibroblasts permits, with modest effort, the induction and isolation of large numbers of newly immortalized cell lines suitable for studying the genetics of mammalian senescence control.

Protocol 2.1

Reagents and Materials

Aseptically prepared
- DMEM15: Dulbecco's modified MEM (DMEM) supplemented with 15% prescreened fetal calf serum.
- Pronase, 0.01% in PBSA
- Trypsin solution, 0.05% in PBSA, supplemented with 1 mg/ml of collagenase 1A.

Protocol

(a) Kill new-born hamsters by neck dislocation and store on ice.
(b) Decapitate and eviscerate the animals, then remove the skins with fine scissors.
(c) Float the skins overnight in 0.01% pronase at 4°C to facilitate separation of the epidermis from the dermis.
(d) Disaggregate isolated dermis by treatment at 37°C with 75 ml of collagenase-trypsin in 100-ml spinner flasks (slow speed). Harvest the supernatant every 20 min. and spin out the cells.
(e) Seed cell suspensions at a density of $1.5 \times 10^7/175$-cm^2 culture flask in 40 ml of DMEM15.
(f) When the cultures are *almost* confluent (i.e., still in log phase), cryopreserve stocks in liquid nitrogen using standard procedures. The cloning efficiency of the recovered cells should be >15% (no feeder layer), otherwise discard the stock. Check samples for the presence of mycoplasma using Hoechst 33258 [Chen, 1977] and/or PCR-based methods.

2.2. Irradiation and Carcinogen Treatment

A single exposure to ionizing radiation or a chemical carcinogen is sufficient to induce immortal variants of SHD cells [Newbold et al., 1982]. The most potent immortalizing agents studied to date are fast neutrons and nickel chloride (Ni^{2+}). X-rays and γ-rays have also proved effective [Trott et al., 1995a].

Protocol 2.2

Reagents and Materials

Sterile materials and reagents
- 75-cm^2, 175-cm^2 flasks
- 50-ml or 125-ml spinner flasks

- DMEM15 (see Section 2.1)
- 0.25 mM NiCl$_2$

Nonsterile equipment
- ^{60}Co source or Van der Graaf accelerator

Protocol

(a) Thaw ampoules of cells and return the contents to 175-cm^2 flasks; subculture (1 : 5 split ratio) after 1–2 days.

(b) For x- and γ-irradiations, trypsinize log phase cultures of passage 3 SHD cells and transfer aliquots of 10^7 cells to 50-ml spinner flasks in 25 ml DMEM15 (x-rays) or 125-ml flasks in 50 ml of medium (γ-rays) and irradiate. Doses reducing cloning efficiency to around 50–60% of the control value (2.5–3 Gy) are the most effective at immortalizing SHD cells. With neutrons (2.3 MeV neutron beam produced by a Van der Graaf accelerator; dose rate, 0.71 Gy/min) irradiate cells as exponentially dividing monolayers in 75-cm^2 flasks.

(c) If nickel is to be used as the immortalizing agent, expose exponentially dividing monolayers to 0.25 mM NiCl$_2$ for 18 h in complete medium (50–60% survival).

2.3. Isolation of Immortal Clones

Protocol 2.3

Reagents and Materials
Aseptically prepared
- 9-cm petri dishes
- Medium: DMEM15
- Materials for clonal isolation and freezing [Freshney, 1994]

Protocol

(a) After treatment, replate cells immediately at 10^6 cells/9-cm dish (10–20 replicates) and passage when just confluent at a fixed split ratio of 1 : 5 initially, reducing to 1 : 3 as the cells begin to enter senescence. Include an equal number of untreated, or solvent-treated, control cultures.

(b) Replace the medium regularly every 3 days.

(c) Immortal variants are seen initially as rare colonies of proliferating cells on a background of senescent counterparts (for photomicrographs depicting examples of emerging immortal SHD clones, see Trott et al., 1995a).

(d) Expand newly immortalized clones to 2 × 10^7 cells and cryopreserve (10 ampoules).

3. CONSTRUCTION OF A HUMAN : RODENT MONOCHROMOSOME HYBRID "DONOR" PANEL

3.1. Tagging Chromosomes in Normal Human Cells Using a Retroviral Vector Incorporating a Mammalian Selectable Marker

In order to be able to transfer human chromosomes individually from normal human cells to newly immortalized hamster cells, and to maintain them in the recipient by drug selection, a panel of monochromosome "donor" hybrids is constructed, each carrying a different human chromosome [Cuthbert et al., 1995]. The first step in this procedure involves tagging chromosomes in a normal human fibroblast cell strain (early passage) with a selectable marker. The use of retroviral vectors for this purpose offers many advantages over plasmid transfection, particularly in terms of the efficiency of marker transfer (which can approach 100% of the infected cells) and in the stability of the resulting tagged chromosomes. We have had a great deal of success using an amphotropic pseudotype of the replication-defective retrovirus vector tgLS(+)HyTK (Fig. 13.1) [see also Lupton et al., 1991]. A producer cell line was constructed [Cuthbert et al., 1995; Trott et al., 1995b] using the packaging cell line PA317 [Miller and Buttimore, 1986], which routinely generates virus titers in excess of 3×10^6 CFU/ml. Producer cell lines should at all times be maintained in selection medium (for the retroviral drug resistance marker) and tested for replication-competent helper virus production by marker rescue assay [Miller and Rosman, 1989]. Only producer lines that routinely test negative for helper virus should be used to infect human cells.

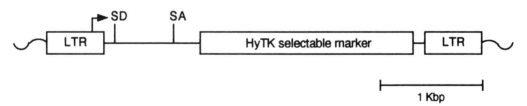

Fig. 13.1. Structure of the tgLS(+)HyTK provirus used to tag human chromosomes with the selectable marker, *Hytk*. The *Hytk* is a fusion gene derived from the bacterial *hph* gene (conferring resistance to hygromycin B) and Herpes simplex virus thymidine kinase (*TK*). Unlike the mammalian protein, HSV *TK* can utilize the acyclic nucleoside Ganciclovir (Syntex Pharmaceuticals) as a substrate. The *Hytk* gene, therefore, may be used as an "in-out" selectable marker; selection for the marker ("in") with hygromycin B-supplemented medium, and selection against ("out") with Ganciclovir.

Protocol 3.1

Reagents and Materials

Aseptically prepared

- 9-cm petri dishes
- Trypsin, 0.25% in PBSA
- Medium: DMEM15
- 50-ml centrifuge tubes
- 0.45-μm sterilizing filters, 25 mm for <50 ml, 47 mm for >50 ml and <500 ml
- Polybrene, 4 mg/ml stock, use at 4 μg/ml
- Hygromycin B, 100 U/ml, in DMEM15
- Freezing medium and materials [Freshney, 1994]

Protocol

(a) Replate virus producing cells in selection medium (9-cm dishes) at a density such that the cultures are 80% confluent after overnight incubation, then replace the medium with 10 ml of drug-free medium.

(b) Replate semi-confluent cultures of early-passage adult human dermal fibroblasts (HDF) at 1 × 10^6 cells/9-cm dish.

(c) After 24 h, collect the virus-conditioned medium (VCM) from the producer cell cultures, centrifuge at 3500 g (5 min) and filter (0.45-μm filter) to remove the cells. Adjust the final concentration of FCS to 15% and add Polybrene (final concentration 4 μg/ml) to the VCM.

(d) Aspirate the medium from the HDF cultures and replace with VCM. Expose the HDF to virus for 24 h before reverting to normal growth medium.

(e) After a 24-h recovery period replate virally infected HDF (1 × 10^6/9-cm dish) in medium containing 800 U/ml hygromycin B.

(f) Pool drug-resistant cells (i.e., those with *Hytk*-tagged chromosomes, see Fig. 13.1) before they reach confluence and cryopreserve for future use in the construction of monochromosomal human : rodent hybrid "donor" panels by direct microcell transfer.

3.2. Optimization of Micronucleus Formation

Exposure of mammalian cells to colcemid for extended periods results in the formation of micronuclei containing the genetic material of a small subset of the chromosome complement (sometimes a single chromosome). The efficiency of micronucleus production and the size of micronuclei produced using a single set of conditions varies dramatically between cell types. For microcell-mediated monochromosome transfer (MMCT), the aim is to maximize the number of micronuclei derived from a single chromosome. Mouse A9 cells respond particularly well to colcemid, producing large numbers of small micronuclei. For this reason A9 cells represent a good choice for constructing human : rodent monochromosome hybrids suitable for use as chromosome donors. In contrast, the response of HDF to colcemid-induced micronucleus formation is relatively poor and the choice of a suitable cell strain will depend on its response to colcemid, which will need to be determined empirically. A careful analysis of the response of selected HDF cell strains at early passage is therefore strongly recommended. Optimal colcemid doses vary between 0.1 and 4.0 μg/ml, with cells being exposed for 48 h in 20% FCS. Visualization and enumeration of HDF micronuclei is best achieved by staining fixed cells with Hoescht 33258 and observing micronuclei with a fluorescence microscope.

Protocol 3.2

Reagents and Materials
Aseptically prepared
- 5-cm petri dishes
- DMEM with 20% FCS
- Colcemid, 100 μg/ml stock

Nonsterile
- Ethanol, 100%
- Hoechst 33258 in Hanks' BSS

Protocol
(a) Replate early-passage (P1-P10) quiescent HDF (previously maintained at confluence for 48 h) at 5×10^5 to 1×10^6 cells/5-cm dish in DMEM containing 20% FCS.
(b) After overnight incubation, add colcemid to partially synchronized HDF at a series of final concentrations ranging from 0.1 to 4.0 μg/ml, and incubate for 48 h.
(c) Fix the cells with 100% ethanol and stain with Hoechst 33258. Enumerate and assess the size of micronuclei under a fluorescence microscope.

3.3. Construction of Monochromosomal Human : Rodent Hybrid "Donor" Panel by Direct Microcell Transfer (Fig. 13.2)

Protocol 3.3

Reagents and Materials
Aseptic materials and reagents
- A9 mouse fibroblasts
- 25-cm^2 flasks (special for centrifugation, see Appendix)

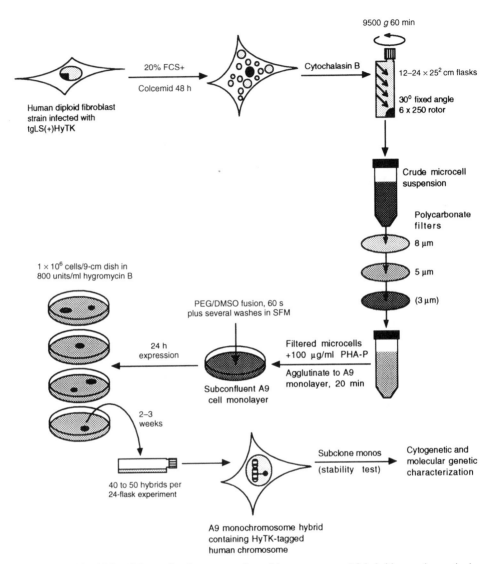

Fig. 13.2. Scheme for the construction of human:mouse-A9 hybrids carrying a single intact human chromosome tagged with the *Hytk* marker.

- 9-cm petri dishes
- DMEM10: DMEM with 10% FCS
- DMEM20: DMEM with 20% FCS
- SFM: serum-free DMEM
- SFM/CB: serum-free DMEM with 5 μg/ml cytochalasin B
- SFM/PHA-P: SFM containing 100 μg/ml phytohaemagglutinin (PHA-P)
- DMEM/Hyg: DMEM10 with 400 U/ml hygromycin B
- Colchicine, 10^{-6} M in SFM

- 8-μm, 5-μm, and 3-μm polycarbonate filters
- 15-ml centrifuge tubes
- SFM/PEG: SFM containing 42.5% PEG (MW 1000) and 8.5% DMSO

Nonsterile equipment
- Accurate two-pan balance
- Centrifuge with programmable acceleration and deceleration
- 30° fixed-angle 6 × 250 ml rotor

Protocol

(a) *Day 1:* Replate quiescent retrovirus-infected HDF cultures into 12 or 24 25-cm^2 flasks in DMEM20.

(b) Add colcemid to a final concentration of 4 μg/ml (for 1BR.2 strain) and incubate for 48 h.

(c) *Day 2:* Replate rodent recipients (usually A9 cells) at 5 × 10^6 cells/9-cm dish 18–24 h before performing microcell fusions. By day 3 the monolayers should be 80–90% confluent.

(d) *Day 3:* Aspirate the colcemid-containing medium from HDF cultures and replace with 30 ml of warm SFM/CB.

(e) Balance pairs of flasks to within 0.02 g prior to centrifugation.

(f) Carefully align the balanced flasks in a 30° fixed-angle 6 × 250 ml rotor (prewarmed to 37°C overnight) such that the growth surface of the flask is lowermost and the caps point towards the axis of the rotor.

(g) Centrifuge at 9500 g for 1 h. (Centrifuges which have fully programmable acceleration and deceleration rates are the best machines for this purpose.) The acceleration phase (lasting 15 min) should be linear up to maximum speed; the deceleration phase should also be linear and of not less than 10-min duration. The overall run time is approximately 85 min.

(h) Pool the crude pellets of microcells from the flasks and filter in series through 8-μm, 5-μm and (in experiments designed to maximize the yield of smaller chromosomes) 3-μm polycarbonate filters.

(i) Centrifuge the filtered microcells at 3500 g for 5 min in 15-ml centrifuge tubes and resuspend in 3 ml of warm SFM/PHA-P.

(j) Wash the A9 cell monolayers (including controls) with three changes of warm SFM before adding the microcell suspension.

(k) Allow the microcells to agglutinate to the 80–90% confluent monolayer for 20 min at 37°C.

(l) Fuse the microcells to the A9 cells by replacing the medium with 3 ml of prewarmed SFM/PEG for 1 min, followed immediately by several thorough washes with warm SFM.

(m) Refeed with DMEM10 and incubate overnight.

(n) *Day 4:* Replate fused cells (and mock-fused cells on control dishes) at 1 × 10⁶ cells/9-cm dish into selection medium (normal growth medium supplemented with 800 U/ml hygromycin B).

(o) *Day 14–24:* Microcell hybrid clones will be large enough for picking after 10–20 days of incubation. Expand clones in DMEM/Hyg for cryopreservation of stocks and for preliminary cytogenetic analysis.

Note: It is essential that hybrid cultures be maintained in selective medium at all times and that they not be allowed to become overconfluent.

4. CHARACTERIZATION OF HYBRIDS

4.1. Preliminary Analysis by G-11 Staining

G-11 staining [Bobrow and Cross, 1974] is a differential chromosome staining procedure that allows one to discriminate between human (pale blue) and mouse A9 (magenta arms with pale blue centromeres) chromosomes in hybrids. Some human chromosomes display characteristic magenta-staining regions, which can in some cases permit tentative identification *per se.* However, the G-11 method will only permit identification of relatively large amounts of human genetic material and cannot be used reliably to assess the purity of candidate monochromosomal hybrids. Use of this valuable and inexpensive technique is therefore best confined to preliminary screening procedures aimed at eliminating hybrids with multiple human chromosomes.

Protocol 4.1

Reagents and Materials
Nonsterile
- Colcemid-arrested cells
- 0.075 M KCl
- 3 : 1 methanol : glacial acetic acid
- Microscope slides
- Coplin jars
- Double distilled water
- Phosphate buffer: 50 mM disodium phosphate adjusted to pH 11.6
- Staining solution (prepare fresh for each batch of slides): To phosphate buffer (pH 11.6, 37°C), add azure B (Sigma) 0.084% (w/v), eosin Y (Sigma) 0.0024% (w/v), and Giemsa (BDH) 1:50(v/v)

Protocol
(a) Prepare metaphase chromosome spreads using standard methods [Mandahl, 1992]. For hypotonic treatment, incubate colcemid-

arrested cells (0.05 μg/ml for 50 min for A9 cells) in 0.075 M KCl and then fix in 3:1 (v/v) methanol:glacial acetic acid (three changes).

(b) Drop fixed cells onto prewashed slides and air dry.

(c) Age slides for 14 days at room temperature prior to G-11 staining.

(d) Preincubate aged slides for 90 s at 37°C in a solution of 50 mM sodium phosphate buffer, pH 11.6.

(e) Place the slides in staining solution. Staining times are dependent on the density of the chromosome spreads but should range between 17 and 25 min. Low-density spreads give the best results.

(f) Remove the slides from the stain (maximum 3 slides/Coplin jar) and quickly rinse in a large excess of double-distilled water.

(g) Mount well-stained preparations.

4.2. Characterization of Hybrids by Fluorescence *In Situ* Hybridization (Chromosome Painting)

A biotin-labelled or digoxigenin-labelled total human DNA probe (Oncor, follow instructions) is employed for conventional ("forward") chromosome painting of human chromosomes in human:mouse A9 microcell hybrids [Cuthbert et al., 1995]. This technique is used to visualize the human DNA content of hybrids, and to establish whether this is associated with a single chromosome or whether hybrids contain rearranged human genetic material not detectable by G-11 staining. The procedure may be repeated after extensive subculturing of selected hybrids and/or recloning to test for stability of the introduced chromosome. Reverse chromosome painting (chromosomal *in situ* suppression hybridization) to normal human metaphases, using probes prepared from *Alu*-PCR-amplified hybrid DNA, is performed to determine the integrity of the introduced human chromosome, and as a sensitive screen for the presence of chimeric human chromosomes [see Cuthbert et al., 1995; also, Cole et al., 1991; Lui et al., 1993].

4.2.1. Preparation of Alu-PCR probes for reverse painting

Protocol 4.2.1

Reagents and Materials

Nonsterile

• (Kit available for this procedure from Boehringer, catalogue # 1636 146)

• High molecular weight (HMW) DNA: Prepare from approximately 1 × 10^7 hybrid cells following standard protocols [Sambrook et al., 1989]

- *Alu* primer ALE 3
- Reaction mixture: 0.5 μg hybrid DNA, 1.2 μM ALE 3 primer, 5 mM MgCl$_2$, 0.5 mM deoxynucleotides (dNTPs) and 4.5 U *Taq* polymerase all in 1$\times$ Promega Thermo DNA polymerase buffer.
- GeneClean
- Tris,10 mM/EDTA, 1 mM, pH 8.0
- 2% agarose in tris/borate EDTA (TBE) buffer
- 2$\times$ SSC: double-strength saline-sodium citrate buffer; 0.3 M NaCl, 30 mM sodium citrate
- SDS, 0.1% in 2$\times$ SSC
- Biotinylated probes purified through Sephadex-G50 columns containing 2$\times$ SSC with 0.1% SDS.
- Sonicated salmon sperm DNA (Sigma)
- Cot1 DNA (BRL)

Protocol

(a) Amplify inter-*Alu* (human) sequences in HMW hybrid DNA with the Alu primer ALE 3. (PCR amplification of sequences in mouse A9 DNA is not detected with this primer). Perform reactions in a total volume of 100 μl of reaction mixture.

(b) After denaturation for 5 min at 95°C followed by 3 min at 80°C (hot start), amplification is performed for 35 cycles: 94°C (1 min), 68°C (1 min), followed by extension at 72°C (5 min). Add Taq polymerase after the initial denaturation step. Monitor each PCR reaction by loading a 10-ml sample onto a 2% agarose gel in TBE buffer.

(c) Purify PCR products with GeneClean, resuspend in Tris/EDTA, pH 8.0, and label by nick translation with biotin-14-dATP (BRL).

(d) Ethanol precipitate the biotinylated probe in the presence of a 100-fold excess of sonicated salmon sperm DNA and a 20-fold excess of Cot1 DNA (BRL), dry, and resuspend in 16 μl of hybridization solution [Carter et al., 1992].

4.2.2. *In situ* hybridization (conventional and reverse painting)

Protocol 4.2.2

Reagents and Materials

Nonsterile

- 2$\times$ SSC (see Reagents and Materials for Protocol 4.2.1)
- 20 μg/ml RNAse in 2$\times$ SSC
- 70% and 100% ethanol on ice
- Acetone
- 70% formamide in 2$\times$ SSC

- DNA probe
- *Alu*-PCR probe

Protocol
(a) Incubate aged slides in 20 μg/ml of RNAse at 37°C for 60 min.
(b) Rinse the slides in 2× SSC (65°C) for 15 min.
(c) Dehydrate through ice-cold 70% (twice), 90% (twice), and 100% ethanol.
(d) Fix in acetone for 10 min.
(e) Denature in a solution of 70% formamide in 2× SSC (65°C, 2 min).
(f) Quench in 70% ethanol.
(g) Dehydrate at room temperature by allowing the denatured slides to air dry.
(h) Denature the probe at 65°C for 10 min and store on ice prior to hybridization. In the case of reverse painting (*Alu*-PCR probe), preanneal the probe to competitor DNA (37°C for 20 min.) before hybridization.
(i) For hybridization, 20 μl of total human DNA probe or 15 μl of *Alu*-PCR probe is used for each denatured slide. Hybridization is performed under sealed coverslips at 37°C, either overnight (total human probe) or for 7 days (*Alu*-PCR probe).

4.2.3. Detection and imaging

Protocol 4.2.3

Reagents and Materials
Nonsterile
- 2× SSC
- 0.1× SSC
- 50% formamide in 2× SSC
- TNFM: 4× SSC containing 0.05% Tween-20 and 5% nonfat milk
- Fluorescein isothiocyanate (FITC)-avidin
- Biotinylated anti-avidin antibody
- 20 ng/ml propidium iodide, or 100 ng/ml DAPI, in Antifade-containing mounting solution

Protocol
(a) Wash the slides twice in 50% formamide (in 2× SSC) followed by one 5-min wash in 2× SSC and one 5-min wash in 0.1× SSC, all at 42°C.
(b) Block in TNFM for 30 min at 37°C.
(c) Detect hybridized biotinylated probe with fluorescein isothio-cyanate–avidin. Amplify the signal by incubating the slides for 20 min with biotinylated anti-avidin antibody (Oncor) in TNFM at room temperature.

(d)   Wash and incubate again with FITC-avidin. (Carry out one round of amplification with the total human DNA probe and two rounds with *Alu*-PCR probes.)

(e)   Counterstain chromosomes with 20 ng/ml of propidium iodide or 100 ng/ml of DAPI in Antifade-containing mounting solution.

(f)   Observe chromosome spreads and capture the digitized images using a confocal laser scanning microscope. (We use a BioRad MRC 600 scanner fitted to a Nikon Optiphot fluorescence microscope; for examples of painted monochromosome hybrids, see Figure 13.3.)

5.   SCREENING THE HUMAN CHROMOSOME COMPLEMENT FOR ANTIPROLIFERATIVE GENES IN HUMAN:RODENT MICROCELL HYBRIDS

5.1.   Chromosome Transfer into Immortalized Syrian Hamster Dermal Cells

The human:rodent monochromosomal hybrids constructed as described above can be used to transfer known human chromosomes singly into any mammalian recipient. The complete human chromosome complement can thereby be screened for senescence-inducing activity. The following procedure, which is a modification of that described above, has been employed with newly immortalized Syrian hamster dermal (SHD) cells as recipients. As discussed above for HDFs, conditions for micronucleation must be opti-

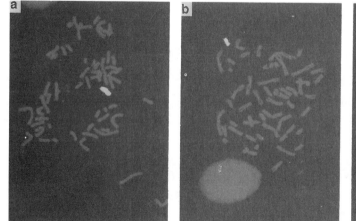
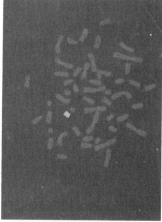

Fig. 13.3.   Examples of monochromosomal human:mouse-A9 hybrid cell lines in which the single human chromosome has been revealed by fluorescence *in situ* hybridization (chromosome painting) using a total human DNA probe. Shown are A9 cell lines carrying human chromosomes X **(a)**, 13 **(b)**, and 19 **(c)**.

Newbold and Cuthbert

mized for each human chromosome "donor" cell line [see also Cuthbert et al., 1995].

Protocol 5.1

Reagents and Materials
Aseptically prepared
- Selection medium: DMEM15 with 600 U/ml of hygromycin B

Protocol
(a) Recipient SHD cell lines are replated at a density of 3×10^6 cells/9-cm dish and incubated overnight (the density is critical and dependent on the cell type). Microcell fusions are performed as described above.

(b) Replate the cells in selection medium at 1×10^6 cells/9-cm dish 24 h after fusion.

(c) Begin to examine hybrid colonies for growth suppression and/or senescence after 1–2 weeks in selection (depending on the growth rate of the recipient cell line). In the case of transfers involving human chromosomes with antiproliferative activity, isolate rare, actively proliferating colonies (potential segregants) after 2 weeks and expand into mass culture for isolation of DNA in preparation for PCR-STS deletion analysis (see Section 6).

The senescence-inducing effects of the human X chromosome may be studied after microcell transfer directly from human diploid fibroblasts into *hprt⁻* derivatives of SHD lines; the *hprt* gene is employed as the selectable marker (conferring resistance to HAT). These experiments are performed using the direct microcell transfer protocol described above. The *hprt⁻* recipient cell lines must be tested for spontaneous reversion to HAT resistance before deciding to use them in experiments. We have isolated two nonrevertible SHD lines for this purpose: 5XH11/6TG4 and 5NL2/6TG1 [Trott et al., 1995a; Cuthbert et al., 1995].

5.2. Selection Protocol

Protocol 5.2

Reagents and Materials
Aseptically prepared
- Complete HAT medium: DMEM15 with hypoxanthine, 10^{-4} M; aminopterin, 10^{-6} M; thymidine 5×10^{-5} M

Protocol

(a) After 24 h of recovery from microcell fusion, replate cells at 5 × 10⁵ cells/9-cm dish in complete HAT medium.

(b) Refeed dishes with fresh selection medium every 3–4 days and, beginning at day 5, monitor colonies for senescence over a 6 to 8-week period.

6. SUBCHROMOSOMAL LOCALIZATION OF CANDIDATE SENESCENCE-INDUCER GENES BY PCR-STS DELETION MAPPING OF HYBRID SEGREGANTS

A wealth of genetically mapped PCR-detectable microsatellite markers (STSs; see Section 1) covering the entire human genome is now available [Gyapay et al., 1994; Chumakov et al., 1995; Hudson et al., 1995]. These markers can be used very successfully to pinpoint the subchromosomal location of antiproliferative and/or senescence genes identified in the above experiments [England et al., 1996; Vojta et al., 1996]. Even with the most powerful antiproliferative response, rare segregants arise the have lost the growth suppressor function. In a high proportion of cases this correlates with nonrandom deletion of human genetic material.

Protocol 6

Reagents and Materials

Nonsterile

• Reaction mixture (per 10 µl):

Hybrid DNA .... 40 ng
Primer .... 1.0 µM
dNTPs, each .... 100 µM
MgCl₂ .... 0.75 mM
Taq polymerase .... 0.2 U
All in 0.5× Promega Thermo DNA polymerase buffer

• 2.5% Agarose
• TBE buffer (see Protocol 4.2)

Protocol

(a) Isolate DNA from putative segregant cell lines using established methods [Sambrook et al., 1989].

(b) Select a set of markers providing good even coverage of the human chromosome being studied (or, in the first instance, a specific region of the chromosome if previous experiments have suggested a candidate location for the senescence gene) and synthesize or purchase PCR primer sets (Research Genetics).

(c) Perform PCR reactions for each STS marker in turn. The following standard set of conditions will usually give good results, but some variations may be required for certain primer sets where amplification of rodent sequences is a problem (uncommon).

(d) Perform reactions in a final volume of 10 μl of reaction mixture.

(e) Denature DNA (94°C for 5 min and 80°C for 8 min) prior to performing PCR amplifications. Use the following PCR conditions: 35 cycles of 94°C (40 s), 56°C (30 s), 72°C (30 s), followed by a final extension cycle of 72°C for 5 min.

(f) Resolve the PCR amplification products on 2.5% agarose gels in TBE buffer.

7. BRIEF SUMMARY OF RESULTS

7.1. Generation of Newly Immortalized SHD Cell Lines

Using the protocols described here, more than 100 immortal SHD cell lines have been generated and cryopreserved as freshly immortalized stocks. A variety of carcinogenic agents have been found to be effective as immortalizing agents, including polycyclic hydrocarbons, their diol epoxide metabolites, aliphatic alkylating agents and ionizing radiation (both high-LET and low-LET). This work, which includes a detailed cytogenetic characterization of the newly immortalized lines, has been published elsewhere [Newbold et al., 1982; Trott et al., 1995a]. The most effective immortalizing agent identified thus far has proved to be soluble nickel ($NiCl_2$). The molecular basis of its superior potency is as yet unclear, although there is some evidence that nickel may cause the functional inactivation of a senescence gene (or genes) via an epigenetic mechanism [Trott et al., 1995a].

7.2. Construction of Human Monochromosome "Donor" Hybrid Panels

The direct microcell transfer approach, when combined with the use of retroviral vectors to tag chromosomes efficiently, has proved to be an extremely effective method of generating human: rodent monochromosome "donor" hybrids, in which the single foreign (human) chromosome is highly stable and transferable intact to a wide variety of recipient cell types. Employing the techniques outlined in this article, a well characterized panel of such hybrids has recently been constructed representing the complete human autosomal complement plus the X chromosome [Cuthbert et al., 1995].

7.3. Mapping Human Antiproliferative Genes

The above human monochromosomal hybrid donor panel has been used to screen the complete human chromosome complement for the presence of antiproliferative (including senescence) genes. During the construction of the panel in the mouse A9 cell line, it was found that two human chromosomes, 1 and 9, were not tolerated by these cells, presumably because they carry cell growth suppressor genes. Consequently, donors for human chromosomes 1 and 9 were isolated using, as recipients, alternative cell lines that proved refractory to the growth suppressive effects. Transfer of individual human chromosomes to newly immortalized SHD cell lines confirmed the antiproliferative properties of chromosomes 1 and 9 and, in addition, has led to the identification of similar activity associated with human chromosomes 19 and X [England et al., 1996]. Furthermore, the cellular effects of introducing the latter human chromosome into SHD lines closely resembled replicative cell senescence. Three of the genes responsible for the induction of growth arrest have been mapped by STS deletion analysis (of rare hybrid segregants) to chromosome 1q (1q25, and possibly 1q42), 9p (9p21), and Xp (tentative). In the case of 9p21, fine deletion mapping of previously uninformative segregants, employing additional markers between D9S162 and D9S171, has provided strong evidence that the proposed tumor suppressor gene $p16^{INK4A}$ (MTS1) is responsible for the arrest of cell growth induced by this human chromosome [England et al., 1996; see also Nobori et al., 1994; Kamb et al., 1994].

8. TELOMERE-DEPENDENT MECHANISMS OF SENESCENCE REGULATION: MAPPING TELOMERASE REPRESSOR GENES IN MICROCELL HYBRIDS DERIVED FROM HUMAN TUMOR CELL LINES

8.1. Background

Convincing evidence has been found for a mechanism involving replication-dependent loss of telomeric DNA that regulates the replicative lifespan of human somatic cells [Counter et al., 1992; Allsopp et al., 1995]. The telomere hypothesis for replicative senescence proposes the existence of a mitotic clock that limits the proliferative potential of somatic cells through continuous loss of telomeric TTAGGG nucleotide repeats at each round of DNA replication [Harley et al., 1990, 1992, 1994]. The telomeres of human germ cells are maintained by the activity of a telomere terminal transferase enzyme, telomerase, which adds TTAGGG repeats to compensate for loss during cell division [Morin, 1989]. It is proposed that in normal somatic cells, reduction in telomere length beyond a critical length provides a signal for entry into irreversible replicative senescence [Shay et al., 1991; Harley et al., 1994]. Telomerase reactivation in immortal somatic cells has been causally implicated in the process of immortalization [Counter et al., 1992, 1994; Greider, 1990].

The twin observations of very short telomeres in many human tumors [Counter et al., 1994; Harley et al., 1994] and a strong correlation between reactivation of telomerase and progression to malignancy in human tumors [Kim et al., 1994; Chadeneau et al., 1995a; Hiyama et al., 1995; Tahara et al., 1995; Hiyama et al., 1996; Li et al., 1996] together suggest that telomere stabilization through telomerase activation is an important event in tumor progression. A correlation between telomerase activation and malignant progression was discovered through the development of a highly sensitive PCR-based assay for mammalian telomerase activity, the Telomere Repeat Amplification Protocol, known as the TRAP assay [Kim et al., 1994]. The emergence of telomerase-positive cells during experimentally induced human cell immortalization *in vitro* has been interpreted as a reflection of events that occur during malignant progression in human tumors [Harley et al., 1994; Shay and Wright, 1996].

Evidence for recessive mutations associated with the emergence of telomerase-positive, immortal human cells has recently emerged from genetic complementation studies [Ohmura et al., 1995; Wright et al., 1996]. They suggest the existence of telomerase repressor genes that regulate a telomere-dependent mechanism of replicative senescence in human cells.

8.2. Scope and Perspective

The availability of a complete panel of human:rodent monochromosomal hybrids for use in functional genetic complementation experiments [Cuthbert et al., 1995] provided our laboratory with the ability to conduct a comprehensive screen of the human genome for putative telomerase repressor genes primarily by chromosome transfer. Subchromosomal mapping of putative telomerase repressor loci is dependent on three factors: (i) use of the TRAP assay to demonstrate telomerase repression in presenescent microcell hybrids that exhibit limited division potential *in vitro* associated with a specific transferred chromosome; (ii) recovery of a panel of revertant immortal hybrids (candidate segregants) that express normal levels of telomerase; and (iii) localization of nonrandom deletions in the exogenous chromosome retained in candidate segregants using PCR-based allelotyping protocols [Louis et al., 1992; Loughran et al., 1997].

The results of recent experiments with cultured rodent cells suggest that a telomere-dependent mechanism for senescence regulation may not exist in mouse and hamster fibroblasts [Chadeneau et al., 1995b; Russo et al., 1998]. This may account for the large discrepancy between rodent and human cells in their propensity towards induced or spontaneous immortalization *in vitro* [Gonos et al., 1992; Newbold et al., 1993; Trott et al., 1995a]. Constitutive expression of telomerase throughout the lifespan of diploid Syrian hamster fibroblasts and in all immortal SHD cell lines tested (Newbold, unpublished results) meant that mapping putative human telomerase repressor genes would not be possible in human:hamster monochromosomal hybrids. Functional genetic complementation studies have therefore been focused on homospecific hybrids derived from human tumor cell lines. The genetic complexity of human:human microcell hybrids may reduce the potential for fine structure subchromosomal mapping of repressor loci in telomerase positive revertants. Unlike heterospecific hybrids, where all human STSs are hemizygous and therefore informative, PCR-mediated analysis of marker deletions on the exogenous chromosome in a human background requires careful screening of markers to ensure each one is potentially informative, i.e., alleles selected for screening against the exogenous chromosome must be distinguishable from the host tumor cell allelotype. Fortunately the ever-increasing density of polymorphic markers placed on linkage and physical maps of the human genome [Gyapay et al., 1994; Chumakov et al., 1995; Hudson et al., 1995] means that the resolution of tumor suppressor and/or telomerase repressor gene mapping in microcell hybrids will continue to increase.

8.3. Summary of Preliminary Results

Preliminary investigations with renal cell carcinoma (RCC) cells revealed the presence of a telomerase repressor locus on chromosome 3 [Ohmura et al., 1995]. Interestingly, allelic loss of heterozygosity (LOH) of polymorphic markers at multiple sites on chromosome 3 has been reported in a variety of human tumors and suggest the presence of multiple tumor suppressor genes. Encouraged by these observations we have conducted studies in tumor cell lines derived from squamous cell carcinomas of the head and neck (SCC-HN) [Edington et al., 1995] and a primary breast carcinoma

[Band et al., 1990]. Having established a strong and exclusive linkage between chromosome 3 transfer and telomerase repression in senescing microcell hybrids derived from these cell lines, panels of immortal, telomerase-positive revertants have now been isolated for STS-PCR deletion analysis with the intention of obtaining a subchromosomal location of the repressor activity. Preliminary results from chromosome 3:SCC-HN revertants have identified a nonrandom region of deletion on the exogenous chromosome at 3p21.2-p21.3 that coincides with LOH data obtained from many cancers, including SCC-HN and breast carcinoma (manuscript in preparation by Andrew P. Cuthbert). Allelotyping of the SCC-HN recipient cell line BICR31 [Edington et al., 1995] also revealed LOH in this region [Andrew P. Cuthbert, unpublished data and see Loughran et al., 1997], suggesting that true genetic complementation of a deleted telomerase repressor locus (or loci) had been achieved.

9. THE FUTURE: MOLECULAR CLONING STRATEGIES FOR SENESCENCE GENES USING APPROACHES BASED ON FUNCTIONAL ASSAYS

Microcell transfer technology has, when used in conjunction with functional assays, proved to be an extremely powerful technique for identifying growth suppressor genes and tumor suppressor genes. In some cases (see Background) the method has permitted the subchromosomal location of the gene to be established. This work has provided a firm foundation for the development of strategies, again based on function, leading to the molecular cloning of novel tumor suppressor genes and senescence genes. The procedure being used in our laboratory has involved the construction of hybrid subpanels of chromosomes of interest (e.g., human chromosomes 1 and 19, and especially X) in which a defined tagged subchromosomal transferable fragment (incorporated into a rodent chromosome) is the only human material present in the rodent cell background. A summary of the protocol for generating such hybrids (known as STFs) is shown in Fig. 13.4 [see also Koi et al., 1993].

Subchromosomal transferrable fragments (STFs; 3–5 Mbp in size) derived from a human monochromosome hybrid with anti-proliferative activity are screened for growth suppressive properties by microcell transfer into the appropriate SHD recipient. STF "donor" hybrids with positive activity can then be used to generate Alu-PCR probes for screening libraries of 650-kb-insert yeast arti-

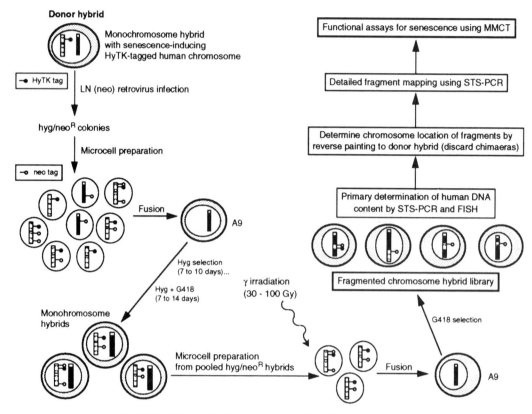

Fig. 13.4. Scheme for the construction of subchromosomal transferrable fragment (STF) hybrids for use in the genetic analysis of cellular senescence. Each STF carries a defined tagged fragment of human genetic material incorporated by recombination into a mouse A9 chromosome.

ficial chromosomes (YACs). YACs selected in this way (retrofitted with selectable markers such as *neo* or *hph*) are then assayed for growth-suppressive properties after transfer to responsive cells by protoplast fusion. This procedure avoids many of the uncertainties and pitfalls associated with positional cloning approaches.

ACKNOWLEDGMENTS

Much of the experimental work leading to the development of the techniques and isolation of hybrids described in this article was supported by grants from the Cancer Research Campaign, the European Commission (Environment Programme), and the Association for International Cancer Research.

REFERENCES

Allsopp RC, Chang E, Kashefi-Aazam M, Rogaev EI, Piatyszek MA, Shay JW, Harley CB (1995): Telomere shortening is associated with cell division in vitro and in vivo. Exp Cell Res 220: 194–200.

Bai L, Mihara K, Kondo Y, Honma M, Namba M (1993): Immortalization of normal human fibroblasts by treatment with 4-nitroquinoline oxide. Int J Cancer 53: 451–456.

Band V, Zajchowski D, Kulesa V, Sager R (1990): Human papilloma virus DNAs immortalize normal human mammary epithelial cells and reduce their growth factor requirements. Proc Natl Acad Sci USA 87: 463–467.

Bobrow M, Cross J (1974): Differential staining of human and mouse chromosomes in interspecific cell hybrids. Nature 251: 77–79.

Carter NP, Ferguson-Smith MA, Perryman MT, Telenius H, Pelmear AH, Leversha MA, Glancy MT, Wood SL, Cook K, Dyson HM, Ferguson-Smith ME, Willatt LR (1992): Reverse chromosome painting: a method for the rapid analysis of aberrant chromosomes in clinical cytogenetics. J Med Genet 29: 299–307.

Chadeneau C, Hay K, Hirte HW, Gallinger S, Bacchetti S (1995a): Telomerase activity associated with aquisition of malignancy in human colorectal cancer. Cancer Res 55: 2533–2536.

Chadeneau C, Siegel P, Harley CB, Muller WJ, Bacchetti S (1995b): Telomerase activity in normal and malignant murine tissues. Oncogene 11: 893–898.

Chen TR (1977): *In situ* detection of mycoplasma contamination in cell cultures by fluorescent Höechst 33258 stain. Exp Cell Res 104: 255–259.

Chumakov IM, Rigault P, Le Gall I, Bellanné-Chantelot C, Billault A, Guillou S, Soularue P, et al. (1995): A YAC contig map of the human genome. Nature 377: 175–297.

Cole CG, Goodfellow PN, Bobrow M, Bentley DR (1991): Generation of novel sequence tagged sites (STSs) from discrete chromosomal regions using Alu-PCR. Genomics 10: 816–826.

Counter CM, Avilion AA, LeFeuvre CE, Stewart NG, Greider CW, Harley CB, Bacchetti S (1992): Telomere shortening associated with chromosome instability is arrested in immortal cells which express telomerase activity. EMBO J 11: 1921–1929.

Counter CM, Hirte HW, Bacchetti S, Harley CB (1994): Telomerase activity in human ovarian carcinoma. Proc Natl Acad Sci USA 91: 2900–2904.

Cuthbert AP, Trott DA, Ekong RM, Jezzard S, England NL, Themis M, Todd CM, Newbold RF (1995): Construction and characterization of a highly stable human:rodent monochromosomal hybrid panel for genetic complementation and genome mapping studies. Cytogenet Cell Genet 71: 68–76.

Edington KG, Louhgran OP, Berry IJ, Parkinson EK: (1995): Cellular immortality: a late event in the progression of human squamous cell carcinoma of the head and neck associated with p53 alteration and a high frequency of allele loss. Mol Carcinog 13: 254–265.

England NL, Cuthbert AP, Trott DA, Jezzard S, Nobori T, Carson DA, Newbold RF (1996): Identification of human tumour suppressor genes by monochromosome transfer: rapid growth-arrest response mapped to 9p21 is mediated solely by the cyclin-D-dependent kinase inhibitor gene, CDKN2A (p16^{INK4A}). Carcinogenesis 17: 1567–1575.

Fan Y-S, Davis L, Shows TB (1990): Mapping small DNA sequences by fluorescence in situ hybridization directly on banded metaphase chromosomes. Proc Natl Acad Sci USA 87: 6223–6227.

Freshney RI (1994): "Culture of Animal Cells, a Manual of Basic Technique." New York: Wiley-Liss, pp 204–206.

Goldstein S (1990): Replicative senescence: the human fibroblast comes of age Science 249: 1129–1133.

Gonos ES, Powell AJ, Jat PS (1992): Human and rodent fibroblasts: model systems for studying senescence and immortalization. Int J Oncol 1: 209.

Greider CW (1990): Telomeres, telomerase and senescence. Bioessays 12: 363–369.

Gyapay G, Morissette J, Vignal A, Dib C, Fizames C, Millasseau P, Marc S, Bernardi G, Lathrop M, Weissenbach J (1994): The 1993–1994 Généthon human genetic linkage map. Nature Genet 7: 246–339.

Harley CB, Futcher AB, Greider C (1990): Telomeres shorten during ageing of human fibroblasts. Nature 345: 458–460.

Harley CB, Kim NW, Prowse KR, Weinrich SL, Hirsch KS, West MD, Bacchetti S, Hirte HW, Counter CM, Greider CW, Piatyszek MA, Wright WE, Shay JW (1994): Telomerase, cell immortality, and cancer. Cold Spring Harb Symp Quant Biol 59: 307–315.

Harley CB, Vaziri H, Counter C, Allsopp RC (1992): The telomere hypothesis of cellular aging. Exp Gerontol 27: 375–382.

Hensler PJ, Annab LA, Barrett JC, Pereira-Smith OM (1994): A gene involved in control of human senescence on human chromosome 1q. Mol Cell Biol 14: 2291–2297.

Hiyama E, Gollahon L, Kataoka T, Kuroi K, Yokoyama T, Gazdar AF, Hiyama K, Piatyszek MA, Shay JW (1996): Telomerase activity in human breast tumours. J Natl Cancer Inst 88: 116–122.

Hiyama E, Yokoyama T, Tatsumoto N, Hiyama K, Imamura Y, Murakami Y, Kodama T, Piatyszek MA, Shay JW, Matsuura Y (1995): Telomerase activity in gastric cancer. Cancer Res 55: 3258–3262.

Hudson TJ, Stein LD, Gerety SS, Ma J, Castle AB, Silva J, et al. (1995): An STS-based map of the human genome. Science 270: 1945–1954.

Kamb A, Gruis NA, Weaver-Fedhouse J, Liu Q, Harshman K, Tavtigan SV, Stockert E, Day RS, Johnson BE, Skolnik MH (1994): A cell cycle regulator potentially involved in genesis of many tumour types. Science 264: 436–440.

Kim NW, Piatyszek MA, Prowse KR, Harley CB, West MD, Ho PLC, Coviello GM, Wright WE, Weinrich SL, Shay JW (1994): Specific association of human telomerase activity with immortal cells and cancer. Science 266: 2011–2015.

Koi M, Johnson LA, Kalikin LM, Little PFR, Nakamura Y, Feinberg AP (1993): Tumor cell growth arrest caused by subchromosomal transferable DNA fragments from chromosome 11. Science 260: 361–364.

Li ZH, Salovaara R, Aaltonen LA, Shibata D (1996): Telomerase activity is commonly detected in hereditary nonpolyposis colorectal cancers. Am J Pathol 148: 1075–1079.

Loughran O, Clark LJ, Bond J, Baker A, Berry IJ, Edington KG, Ly I-S, Simmons R, Haw R, Black DM, Newbold RF, Parkinson EK (1997): Evidence for the inactivation of multiple replicative lifespan genes in immortal human squamous cell carcinoma keratinocytes. Oncogene 14: 1955–1964.

Louis DN, von Deimling A, Seizinger BR (1992): A $(CA)_n$ dinucleotide repeat assay for evaluating loss of allelic heterozygosity in small and archival human brain tumour specimens. Am J Pathol 141: 777–782.

Lui P, Siciliano J, Seong D, Craig J, Zhao Y, de Jong PJ, Siciliano MJ (1993): Dual Alu polymerase chain reaction primers and conditions for isolation of human chromosome painting probes from hybrid cells. Cancer Genet Cytogenet 65: 93–99.

Lupton SD, Brunton LL, Kalberg VA, Overell, RW (1991): Dominant positive and negative selection using a hygromycin phosphotransferase-thymidine kinase fusion gene. Mol Cell Biol 11: 3374–3378.

Mandahl N (1992): Methods in solid tumour cytogenetics. In Rooney DE, Czepulkowski BH (eds): "Human Cytogenetics, a Practical Approach, Vol 2." Oxford: IRL Press.

Miller AD, Buttimore C (1986): Redesign of retrovirus packaging cell lines to avoid recombination leading to helper virus production. Mol Cell Biol 6: 2895–2902.

Miller AD, Rosman GJ (1989): Improved retroviral vectors for gene transfer and expression. Biotechniques 7: 980–990.

Morin GB (1989): The human telomere terminal transferase enzyme is a ribonucleoprotein that synthesizes TTAGGG repeats. Cell 59: 521–529.

Newbold RF (1985): Multistep malignant transformation of mammalian cells by carcinogens: induction of immortality as a key event. In Barrett JC, Tennant RW (eds): "Carcinogenesis: A Comprehensive Survey, Vol 9." New York: Raven Press, pp 17–28.

Newbold RF, Cuthbert AP, Themis M, Trott DA, Blair AL, Li W (1993): Cell immortalization as a key, rate-limiting event in malignant transformation: approaches toward a molecular genetic analysis. Toxicol Lett 67: 211–230.

Newbold RF, Overell RW (1983): Fibroblast immortality is a prerequisite for transformation by EJ c-Ha-ras oncogene. Nature 304: 648–651.

Newbold RF, Overell RW, Connell JR (1982): Induction of immortality is an early event in malignant transformation of mammalian cells by carcinogens. Nature 299: 633–635.

Nobori T, Miura K, Wu DJ, Lois A, Takabayashi K, Carson DA (1994): Deletions of the cyclin-dependent kinase-4 inhibitor gene in multiple human cancers. Nature 368: 753–756.

Ohmura H, Tahara H, Suzuki M, Ide T, Shimizu M, Yoshida MA, Tahara E, Shay JW, Barrett JC, Oshimura M (1995): Restoration of the cellular senescence program and repression of telomerase by human chromosome 3. Jpn J Cancer Res 86: 899–904.

Russo I, Silver AJR, Cuthbert AP, Newbold RF (1998): Novel telomere-independent mechanism of senescence is the sole barrier to immortalization in rodent cells. (submitted for publication).

Sambrook J, Fritsch EF, Maniatis T (1989): "Molecular Cloning, a Laboratory Manual, 2nd ed." Cold Spring Harbor, NY: Cold Spring Harbor Laboratory Press, 3 vols.

Saxon PJ, Stanbridge EJ (1987): Transfer and selective retention of single specific human chromosomes via microcell-mediated chromosome transfer. Methods Enzymol 151: 313–325.

Shay JW, Wright WE (1996): The reactivation of telomerase activity in cancer progression. Trends Genet 12: 129–131.

Shay JW, Wright WE, Werbin H (1991): Defining the molecular mechanisms of human cell immortalization. Biochim Biophys Acta 1072: 1–7.

Tahara H, Kuniyasu H, Yokozaki H, Yasui W, Shay JW, Ide T, Tahara E (1995): Telomerase activity in preneoplastic and neoplastic gastric colorectal lesions. Clin Cancer Res 1: 1245–1251.

Trott DA, Cuthbert AP, Overell RW, Russo I, Newbold RF (1995a): Mechanisms involved in the immortalization of mammalian cell by ionizing radiation and chemical carcinogens. Carcinogenesis 16: 193–204.

Trott DA, Cuthbert AP, Todd CM, Newbold RF (1995b): Novel use of a selectable fusion gene as an "in-out" marker for studying genetic loss in mammalian cells. Mol Carcinog 12: 213–224.

Vojta PJ, Futreal PA, Annab LA, Kato H, Pereira-Smith OM, Barrett JC (1996): Evidence for two senescence loci on human chromosome 1. Genes Chromosomes Cancer 16: 55–63.

Wright WE, Brasiskytr D, Piatyszek MA, Shay JW (1996): Experimental elongation of telomeres extends the lifespan of immortal × normal cell hybrids.
EMBO J 15: 1734–1741.

APPENDIX: MATERIALS AND SUPPLIERS

Materials	Suppliers
Agarose	Gibco BRL
Aminopterin	Sigma
Antibiotics	Gibco BRL
Antifade-containing mounting solution	Vector Laboratories
Azure B	Sigma
Biotin conjugated anti-avidin antibody	Oncor, Vector
Biotin or digoxigenin-labelled total human DNA probe	Oncor, Vector
Centrifugation flasks, 25 cm^2	Nunclon
Colcemid	Sigma
Collagenase, grade 1A	Sigma
Cot 1 DNA	Gibco BRL
Cytochalasin B	Sigma
Dimethyl sulfoxide (DMSO)	Sigma
Dulbecco's modified Eagle's medium (DMEM)	Gibco BRL
Eosin Y	Sigma
Fetal calf serum	Gibco BRL
Fluorescein isothiocyante-avidin	Oncor, Vector
Formamide	Sigma
Ganciclovir (under licenced agreement)	Syntex
Geneticin G418	Gibco BRL
Giemsa stain	Fisher
Höescht 33258 (bizbenzimide)	Sigma
Hygromycin B	Calbiochem
Hypoxanthine	Sigma
Mycoplasma PCR kit	Gen-Probe
Newborn calf serum	Gibco BRL
NiCl$_2$	Sigma
PCR primer sets	Research Genetics
Phytohemagglutinin PHA-P	Sigma
Polybrene (hexadimethrine bromide)	Sigma
Polycarbonate filters and holders	Nuclepore (Corning Costar)
Polyethylene glycol (PEG), mw 1000	Sigma
Pronase	Sigma
Propidium iodide	Sigma
Salmon sperm DNA	Sigma
Taq polymerse	Promega or Perkin Elmer
Thermo DNA polymerase buffer	Promega
Thymidine	Sigma
Tissue culture plastics	Nunclon
Trypsin (bovine pancreas, TRL)	Worthington Enzymes
Tunicamycin	Calbiochem

14

Methotrexate and DHFR Amplification

Michael H. Ricketts

Department of Psychiatry, UMDNJ-Robert Wood Johnson Medical School, Piscataway, New Jersey 08854

DNA Transfer to Cultured Cells, Edited by Katya Ravid and R. Ian Freshney.
ISBN 0-471-16572-7 © 1998 Wiley-Liss, Inc.

I. INTRODUCTION

Investigations directed at understanding how cancer cells become resistant to chemotherapeutic agents, such as methotrexate, led to the understanding that gene amplification is ubiquitous. Exposure of cells to methotrexate will result in the selective survival of those cells with increased dihydrofolate reductase (DHFR) activity, most frequently as a result of DHFR gene amplification [Alt et al., 1978; Schimke, 1988]. A number of investigators have subsequently shown that DNA sequences co-transfected with a DHFR expression vector are co-amplified when the cells are subjected to selection in methotrexate [Wigler et al., 1980; Kaufman and Sharp, 1982; Ringold et al., 1982; Christman et al., 1982]. It is now known that, in addition to DHFR, a number of other genes can be used as markers to amplify expression of a co-transfected gene of interest [reviewed by Kaufman, 1990a]. This chapter will focus primarily on the methods for co-amplification of genes by methotrexate selection in Rat-1 cells.

The term "amplification by methotrexate selection", while commonly used, should not be interpreted to mean that methotrexate (or any other "amplifiable" marker) is effecting DNA amplification. DNA amplification is ubiquitous, and methotrexate functions to selectively kill cells in which the DHFR gene has not become amplified.

The methods presented should enable investigators to express a protein of interest at high level in cultured mammalian cells. The principle is to introduce a vector expressing a target protein together with vectors to enable selection and amplification of all the transfected DNA. The transfected DNA vectors should integrate into the host cell DNA as a unit, so that amplification of the introduced DHFR gene by methotrexate selection will lead to co-amplification of the gene of interest. This co-integration, enabling amplification of a linked gene, has been termed cotransformation [Wigler et al., 1977]. The transfection and amplification of an expression vector in Rat-1 cells, from the initial transfection until clones with high expression levels are obtained after a few steps of methotrexate selection, can be expected to take about 4 months.

The production of cell lines expressing a high level of a protein of interest has a number of important applications. These include: (i) characterization of the effect of high level expression on cell physiology or parameters of transformation, such as the effects and interactions of oncogenes and tumor suppressor genes [Hudziak et al., 1987; Ricketts and Levinson, 1988; Cohick and Clemmons,

1994], (ii) establishment of a constant source of protein for functional studies after isolation, or in cells or cellular fractions [Hussain et al., 1992; Ricketts et al., 1992; Isola et al., 1995], and (iii) production of a protein to use as a reagent for research or therapeutic application. This is particularly important where the post-translational processing of the protein of interest cannot be correctly completed in prokaryotic cells. By introducing specific mutations into the expression vector the effect of these mutations on the function of the protein of interest can be investigated. The post-translational processing and targeting of proteins can also be more easily studied if cell lines expressing high levels of the protein are established [Israel et al., 1992].

2. PREPARATION OF REAGENTS AND MEDIA

2.1. Choice of Cell Lines

In theory, any cell line that can grow as an isolated colony can be used to obtain stable cell lines for amplification of transfected DNA by methotrexate selection. It is not necessary that the cell line be deficient in DHFR activity. Examples of cell lines that have been successfully used to amplify heterologous gene products by methotrexate selection include Rat-1 cells [Ricketts and Levinson, 1988; Ricketts et al., 1992], NIH 3T3 cells [Hudziak et al., 1987; Isola et al., 1995], Chinese hamster ovary (CHO) cells [Israel et al., 1992; Stromqvist et al., 1994; McCartney et al., 1995], a Chinese hamster lung cell line [Hussain et al., 1992], and a rat liver cell line [Jou et al., 1993]. However, the stability of amplified DNA in the absence of continued selection is influenced by the choice of cell line. DHFR genes amplified in mouse cell lines are usually associated with extrachromosomal elements (double minute chromosomes) and are, consequently, less stable in the absence of selection [Murray et al., 1983]. The expression levels of heterologous proteins in Rat-1 cells and in CHO cells have been found to remain high in the absence of selection [Kaufman et al., 1983; Ricketts et al., 1991]. The primary methods described in this chapter are those that have been successfully used to obtain high-level expression of heterologous genes in Rat-1 cells. The same methods are generally applicable to other cell types, although the published literature should be consulted to note specific methodological differences. Some of the methods for DNA transfection and amplification using DHFR-deficient CHO cells [Kaufman, 1990a] are presented in the section on alternative methods.

2.2. DNA Vectors

Successful amplification of a gene of interest can be achieved by co-transfecting cells with separate plasmid vectors for primary selection (G418 resistance), amplification (DHFR), and expression of the protein of interest. This approach, using three separate vectors, has the advantage that the relative amount of the selection and target genes can be varied in the transfection. A useful and commonly used vector for primary selection is pSV2Neo [Southern and Berg, 1982], which confers resistance to Geneticin in mammalian cells. For an amplifiable vector, pFD11, which directs the expression of murine DHFR, can be used [Simonsen and Levinson, 1983; Ricketts and Levinson, 1988], as can the alternative DHFR-encoding vectors pCVSVEII-DHFR, pAdD26SV(A), and pED [Ausubel et al., 1996; Kaufman, 1990b; Kaufman et al., 1991]. The vector directing expression of the target gene should be constructed with a suitable promoter, such as the SV40 early promoter, and should have appropriate RNA cleavage and polyadenylation signal sequences. A number of suitable vectors for insertion of sequences encoding a specific gene of interest are available commercially. For an overview of mammalian expression vectors see Kaufman [1990b]. Many expression vectors contain a neomycin resistance gene for primary selection on the same plasmid as the target gene, so that a single vector can be used to direct expression of neomycin resistance and the gene of interest. A unit directing the expression of DHFR could also be included in the same vector.

The DNA vectors are plasmid-based expression vectors and the plasmid DNA should be prepared by cesium chloride centrifugation [Ausubel et al., 1996] or with a Qiagen plasmid isolation kit (Qiagen Inc., #12143) to obtain pure covalently closed circular plasmid DNA for the transfection of cultured cells. Ensure that the plasmid preparations are sterile by air drying in a tissue culture hood after the final 70% ethanol wash and dissolving the plasmid DNA in sterile water.

2.3. Media and Reagents

Calcium chloride for transfection (2.5 M CaCl$_2$)

$CaCl_2 \cdot 2H_2O$ .... 36.75 g
Water to .... 100 ml

Sterilize through 0.22-μm nitrocellulose filter.
Store aliquots at $-20°C$

Dialyzed fetal bovine serum

This can be obtained commercially (e.g., Life Technologies, catalogue # 10440-014). Alternatively, fetal bovine serum can be dialyzed against PBSA or Hanks' balanced salt solution, using a membrane with a cut-off of about 10,000 d and filter-sterilized through a 0.2-μm membrane [Freshney, 1994].

G418 (Geneticin), 100×

Prepare a 40 mg/ml stock of Geneticin in 1× HEPES-buffered saline. (*Note:* Use the mass of active G418 to determine the concentration; this is indicated on the product and is generally between 500 and 800 mg/g). Adjust the pH of the solution to 7.2 with 5 M NaOH before adjusting the final volume. Filter (0.22 μm) to sterilize and store in aliquots at $-20°$C.

G418 medium

G418 ...... 0.4 mg/ml

in DMEM : F12/FB

HEPES-buffered saline, 2×

A solution of 280 mM NaCl, 50 mM HEPES, 1.5 mM Na_2HPO_4, pH 7.1 is made as follows:

NaCl .... 16.4 g
HEPES acid ...... 11.9 g
NaH_2PO_4 .... 0.21 g

Add distilled water to about 800 ml, titrate with 5 M NaOH to pH 7.1, and bring the final volume to 1 liter. Filter-sterilize the solution, aliquot and store at $-70°$C. Before use, thaw the aliquot completely and mix well.

Glycerol in PBSA, 20%

A 20% (v/v) glycerol stock in phosphate-buffered saline for glycerol shock treatment of transfected cells is made up as follows:

PBSA 10× .... 20 ml
Glycerol .... 20 ml
H_2O .... 160 ml

Autoclave the solution to sterilize.

Methotrexate

Prepare a 5 mM methotrexate stock solution by dissolving 0.1136 g of (+)amethopterin in 50 ml of F12 : DMEM (without glycine, hypoxanthine, and thymidine). Filter-sterilize the solution and store protected from light at $-20°$C. Working solutions can be made from this stock solution.

Methotrexate selection medium

F12:DMEM (1:1) without glycine,
hypoxanthine, and thymidine .... 500 ml
Dialyzed fetal bovine serum (final
concentration 7%) .... 37.5 ml
100× stock of penicillin streptomycin .... 5 ml

Add methotrexate to obtain the desired concentration for selection.

PBSA (phosphate buffered saline without calcium and magnesium), 10×

A 10× solution to give a working solution of 137 mM NaCl, 2.7 mM KCl, 4.3 mM $Na_2HPO_4 \cdot 7H_2O$, and 1.4 mM KH_2PO_4, with pH~7.3, is made up as follows:

NaCl .... 80 g
KCl .... 2 g
$Na_2HPO_4 \cdot 7H_2O$ .... 11.5 g
KH_2PO_4 .... 2 g

Dissolve in a final volume of 1 liter water and sterilize by autoclaving.

Rat-1 cell culture medium (F12:DMEM/FB)

F12:DMEM (1:1)
Penicillin (100 U/ml) and streptomycin (100 (μg/ml)
10% fetal bovine serum

Complete ADT medium

Adenosine .... 10 μg/ml
Deoxyadenosine .... 10 μg/ml
Thymidine .... 10 μg/ml
Fetal bovine serum, heat inactivated 10%

In alpha medium without nucleosides (Life Technologies, catalogue #32561)

Complete alpha medium

Fetal bovine serum, heat inactivated 10%

In alpha medium without nucleosides (Life Technologies, catalogue #32561).

TE buffer

Tris .... 10 mM
EDTA .... 1 mM

Make this buffer from stock solutions of Tris and EDTA with pH 8.0.

Trypsin/EDTA

0.25% trypsin, 1 mM EDTA in Hanks' balanced salt solution or PBSA

3. TRANSFECTION AND SELECTION OF CELLS

3.1. Preparation of Rat-1 Cells for Transfection

Protocol 3.1

Reagents and Materials

Aseptically prepared

- Rat-1 cells (available from author)
- Plasmid vectors and controls
- Culture medium (F12 : DMEM/FB)
- Petri dishes, 60 mm

Protocol

(a) Propagate Rat-1 cells in Rat-1 cell culture medium at 37°C in a humidified 5% CO_2 atmosphere. Maintain stock cultures at subconfluent densities by splitting the cells 1 : 30 approximately every 4–7 days.

(b) Prepare Rat-1 cells 24 h before the transfection by plating about 5×10^5 cells per 60-mm dish to be transfected. This number of cells can be estimated by splitting a subconfluent 100-mm dish to ten 60-mm dishes. The dishes should be about 50% confluent at the time of transfection. It is recommended that the transfections be done in duplicate, and that appropriate controls, omitting specific vectors, be included.

(c) Allow the cells to attach for a few hours, aspirate off the medium, and add 5 ml of fresh medium to the cells.

(d) Incubate the cells overnight at 37°C in a 5% CO_2 atmosphere.

3.2. Calcium Phosphate/DNA Transfection

3.2. Calcium Phosphate/DNA Transfection

Protocol 3.2

Reagents and Materials

Aseptically prepared

- Plastic tubes, clear, 5 ml (12 × 75 mm)
- Expression vector encoding the gene of interest (or control plasmid DNA)
- Spiked polypropylene tube holder

- F12:DMEM/10FB: F12:DMEM medium with 10% FBS
- F12:DMEM/SF: F12:DMEM without serum

Protocol

(a)   In clear 5 ml sterile tubes add and mix:

Expression vector .... 5 μg
pFD11 (encoding DHFR)(available from author) .... 0.5 μg
pSV2Neo (encoding G418 resistance) .... 0.5 μg
Sterile water to a final volume of .... 225 μl
CaCl$_2$, 2.5 M .... 25 μl

(b)   Slowly add 250 μl 2× HEPES buffered saline, with continual gentle mixing. The mixing can be achieved by moving the tube across an uneven surface (such as the tube holder, or by gently bubbling filtered air through the solution, while slowly adding the 2× HEPES from a pipette.

(c)   Incubate the solution for 30 min at room temperature. A very fine precipitate should be visible in the tubes, particularly when viewed against the light and compared with a tube containing 500 μl of clear solution.

(d)   Remove the Rat-1 cells from the incubator, distribute the DNA/calcium phosphate solution over the cells and agitate each dish gently to mix. Return the cells to the incubator without undue delay.

(e)   Incubate the cells for 5–6 h before proceeding with the glycerol shock (steps f–i). The glycerol shock improves the transfection efficiency.

(f)   Remove only a few dishes of transfected cells from the incubator at a time and aspirate the medium from the cells.

(g)   Add 1 ml of 20% glycerol in PBSA to each dish. Do this carefully to avoid washing the cells off part of the dish. Allow the solution to distribute over the cells by carefully tilting the dish.

(h)   Incubate for 45 s, then carefully add 4 ml of F12:DMEM/SF and mix by swirling. It is important to dilute the glycerol on the cells in a timely fashion as the glycerol is toxic to the cells.

(i)   Aspirate the medium, rinse the cells with another 5 ml of F12:DMEM/SF, add 5 ml of F12:DMEM/10FB, and return the cells to the incubator.

3.3.   Geneticin Selection

Protocol 3.3

Reagents and Materials
Aseptically prepared
- Trypsin/EDTA
- PBSA

- G418 medium
- Petri dishes, 100 mm

Protocol

(a) After incubation of the transfected cells for 24–48 h, rinse the cells with 3–4 ml of PBSA and use 1.5 ml of trypsin to split the cells from each 60-mm dish to two 100-mm dishes containing G418 medium.

(b) Replace the medium with fresh medium containing G418 every 3–4 days. After a few days the cells will die and detach from the dish. Colonies of G418-resistant cells should become visible after 10 days. View the plate against the light at an angle from below to see the colonies as opaque patches. The colonies can be circled with a marker pen to facilitate their location for microscopic examination, and to help isolate colonies with cloning cylinders.

(c) When the colonies reach 2–3 mm in diameter, individual colonies can be isolated and expanded using cloning cylinders (see Section 3.5) and analyzed for expression from the transfected gene.

After expansion, individual colonies can be used for amplification by methotrexate selection, but it is recommended that the pooled clones obtained after Geneticin selection be used for methotrexate selection (see Fig. 14.1).

(d) To pool the G418-resistant colonies aspirate the medium from the dish and rinse the cells with 3–5 ml of PBSA. Use 4 ml of trypsin/EDTA to loosen the cells. During trypsinization it is useful to mark the colony positions under the plate for counting, as they become very opaque and are easily visible during this step.

3.4. Amplification by Methotrexate Selection

Protocol 3.4

Reagents and Materials

Aseptically prepared
- Transfected cells
- Methotrexate selection medium (see Section 2.3), 70 nM and 200 nM
- Petri dishes, 100 mm

Protocol

(a) Plate 2×10^5 cells in 10 ml of methotrexate selection medium in each 100-mm dish. The use of at least two concentrations of methotrexate at each selection step, e.g. 70 nM and 200 nM, is recommended.

(b) Incubate the cells at 37°C in a humidified 5% CO_2 atmosphere, and replace the medium every 3–4 days, maintaining the same methotrexate concentration. Colonies of methotrexate-resistant cells should be visible after 2 weeks.

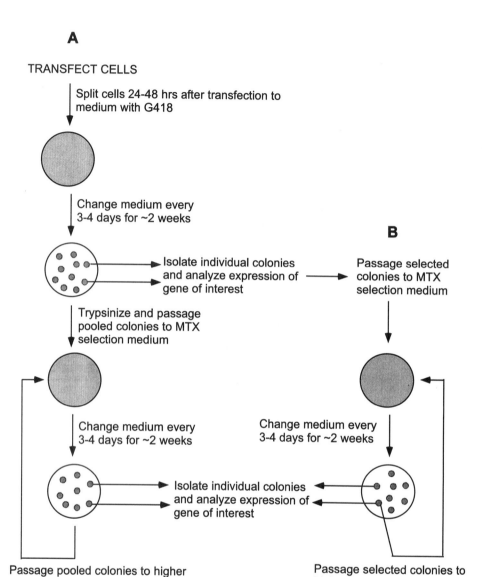

A

TRANSFECT CELLS

Split cells 24-48 hrs after transfection to medium with G418

Change medium every 3-4 days for ~2 weeks

Isolate individual colonies and analyze expression of gene of interest

Trypsinize and passage pooled colonies to MTX selection medium

Change medium every 3-4 days for ~2 weeks

Isolate individual colonies and analyze expression of gene of interest

Passage pooled colonies to higher level of MTX

B

Passage selected colonies to MTX selection medium

Change medium every 3-4 days for ~2 weeks

Passage selected colonies to higher level of MTX

Fig. 14.1. Scheme for transfection and selection of Rat-1 cells to obtain cells expressing high levels of heterologous protein. **A.** Procedure for methotrexate selection using the whole population of cells obtained at previous level of selection. **B.** Alternative procedure, using individual characterized clones to select for amplified DNA at higher levels of methotrexate. MTX = methotrexate.

(c) Use cloning cylinders to isolate and expand colonies for the analysis of expression levels (see Sect. 3.5 below).

(d) Pool the remaining colonies from a dish and passage 2×10^5 cells per 100-mm dish containing higher concentrations of methotrexate in methotrexate selection medium.

For example, passage colonies selected at 70 nM methotrexate to 250 nM and 700 nM methotrexate. Avoid using very large steps as this leads to selection of colonies resistant to methotrexate by mechanisms other than DNA amplification.

(e) Continue with selection and analysis of expression in individual colonies as illustrated in Fig. 14.1.

3.5. Preparation and Use of Cloning Cylinders

Protocol 3.5

Reagents and Materials
• Trypsin/EDTA
• Cloning cylinders (Bellco), prepared for use by autoclaving in a covered glass dish with sufficient petroleum jelly to cover the bottom of the dish to a depth of 1–2 mm
• Pasteur pipettes

Protocol
(a) Use bent forceps (sterilized by ethanol and flaming) to place the cloning cylinder over the colony of interest, the position of which has been marked on the underside of the dish. Make sure the cylinder is firmly seated around the colony.

(b) Carefully aspirate the medium inside the cylinder.

(c) Add trypsin/EDTA to fill about 75% of the cylinder, and allow 2–3 min for the cells to detach. A rinse with a few drops of PBSA before trypsinization is optional with Rat-1 cells, but important with cells that adhere very tightly.

(d) With a sterile Pasteur pipette and bulb, pipette the solution up and down a few times to dislodge the cells, and transfer the cells to a 35-mm or 60-mm dish with the appropriate selection medium for propagation of the colony. Use a microscope to verify that the colony is no longer in its original position and that the dispersed cells are in the recipient dish.

4. ANALYSIS OF HETEROLOGOUS PROTEIN EXPRESSION

At each stage of selection the level of expression of the gene of interest can be determined in selected colonies and in pooled cell populations. Southern blots can be used to monitor the extent of target gene amplification, and Northern blot analysis can be used to determine the level of mRNA. However, if possible, it is preferable to monitor directly the level of expression of the protein of interest. Depending on the nature of the protein and the availability of antibodies, one or more of the following approaches may be useful.

If the expressed protein has a biological activity that can be determined in cell extracts of the isolated clones, this can be exploited. The expression level of an enzyme can be determined in cell extracts using standard assays for the particular enzyme being expressed. Details of the procedure will depend upon the specific protein being expressed. If an antibody directed against the protein of interest is available, expression can be detected and quantified by either Western blot analysis or metabolic labeling with radioactive amino acids and immunoprecipitation, followed by gel electrophoresis and autoradiography.

The stability of the transfected DNA can be monitored by removing the cells from methotrexate selection and analyzing DNA (by Southern blot) or expression of the heterologous protein in cells cultured for weeks or months without selection.

5. ADVICE ON SAFETY PROCEDURES

Methotrexate is a known carcinogen and care should be taken to handle methotrexate with gloves and in a fume hood.

6. ALTERNATIVE METHODS

In deciding whether the above methods are suited to one's goals, a broad range of alternative approaches should be considered. First, would expression in nonmammalian systems, such as *E. coli* or insect cells be suitable? If mammalian cells are important (e.g., to process correctly the expressed protein), would transient transfection experiments be suitable? Transient transfection studies are certainly less time-consuming, but do not provide a constant supply of high-level expression of the protein of interest.

Detailed methods have been presented for obtaining high-level expression in Rat-1 cells. Alternative cell lines can be used to obtain high levels of expression by methotrexate selection. In particular, CHO cells offer a number of advantages if their transformed condition is not an obstacle to the final aims. A protocol for the transfection and DHFR selection of CHO cells is provided below. Rat-2 cells, a thymidine kinase–deficient derivative of Rat-1, can also be used [Tropp, 1981]. Rat-2 cells, available from American Type Culture Collection (ATCC) are not transformed and are efficiently transfected.

6.1. Transfection and DHFR Selection of CHO Cells

CHO cell lines deficient in DHFR are available [Urlaub and Chasin, 1980], so that methotrexate, or a nucleoside-deficient medium, can be used for primary selection. Amplified genes are generally stable in CHO cells, and the cells can be grown attached or in suspension. The methods outlined here for amplification of heterologous DNA sequences in DHFR deficient CHO cells can be found in more detail in Kaufman [1990a] and Ausubel et al. [1996, Unit 16.14]. CHO cell lines that can be used are CHO DXB11 or CHO DG44 (available from Lawrence Chasin, Columbia University), or CHO GRA (available from Randal Kaufman, Genetics Institute, Cambridge, MA, USA).

Protocol 6.1

Reagents and Materials
- Complete ADT medium
- Plasmid DNA encoding DHFR and the protein of interest
- 10% glycerol in PBSA
- Complete alpha medium (Life Technologies, catalogue #32561)

Protocol
(a) Culture the CHO cells in complete ADT medium.
(b) On the day before transfection, split a confluent dish of CHO cells 1:15 in complete ADT medium.
(c) Transfect the cells with 5–10 μg of plasmid DNA encoding DHFR and the protein of interest. The recommended method is calcium phosphate precipitation (see Section 3.2), followed about 6 h later by a 3-min glycerol shock (with 10% glycerol in PBSA).
(d) Allow about 2 days for the cells to reach confluence, then split them 1:15 into complete alpha medium for selection. Cells need DHFR to synthesize necessary nucleosides not available in this selective medium. Methotrexate is not needed for the selection.
(e) After incubation for 10–14 days, stable colonies can be marked and isolated with cloning cylinders (see Section 3.5) and expanded. As with Rat-1 cells, individual colonies or pooled colonies can be used for amplification in methotrexate. It is prudent to establish that the selected cells do express the transfected gene of interest before proceeding with amplification by methotrexate selection.

(f) Split a confluent dish of transfected cells 1:6 into complete alpha medium containing methotrexate at three concentrations, e.g., 5 nM, 20 nM, and 50 nM. This will select for cells with increased levels of DHFR, but because of endogenous nucleoside reserves the cells will initially continue to grow rapidly.

(g) Whenever the cells reach confluence split them again into complete alpha medium with methotrexate until they begin to assume a normal morphology and grow rapidly. Then increase the dilution of cells in a stepwise manner to 1:8, 1:10, and 1:15, all in methotrexate. When they reach confluence only 3 days after a 1:15 split they are ready for selection at a higher level of methotrexate.

(h) Increase the concentration of methotrexate about fourfold and repeat the procedure for the next amplification step. Each amplification step takes 3–4 weeks. Selection above a methotrexate concentration of 80 μM does not result in further amplification, as the transport of methotrexate into the CHO cell becomes limiting.

An alternative approach to amplification with methotrexate is to expose the CHO cells initially to higher methotrexate levels (e.g., 20 nM and 80 nM) and to isolate individual colonies after about 10 days to characterize the expression of the target protein. After expansion of individual colonies, the cells are subjected to further selection at 16-fold and 64-fold the level of methotrexate at which they were originally selected [Kaufman, 1990a; Ausubel et al., 1996].

6.2. Alternative Transfection Methods

Alternatives to the calcium phosphate transfection method can be used to introduce DNA into cells. This will be particularly important if the chosen cell line is not efficiently transfected using the calcium phosphate method. Alternative methods include electroporation (see Chapters 3, 4, and 5) and liposome-mediated transfection (see Chapter 10). These and other methods are described in Ausubel et al. [1996].

6.3. Alternative Selection Vectors

Hygromycin B can also be used as a primary selection agent in conjunction with an expression vector encoding hygromycin B phosphotransferase. The concentration required for selection depends upon the cell line. Rat-1 clones obtained after selection with G418 and methotrexate can be subsequently co-transfected with a second gene of interest and a hygromycin B-resistance gene; the colonies can then be selected using 0.4 mg/ml hygromycin B [Ricketts and Levinson, 1988]. This provides a means of introducing a second gene of interest into a cell line already expressing high levels of the first target gene.

Methotrexate can be used for primary selection in cells that are not deficient in DHFR using a plasmid encoding a mutated DHFR. This has been done in NIH 3T3 cells, using the DHFR gene from an A29 cell line derivative [Christman et al., 1982]. The initial selection was at 440 nM methotrexate. However, in the murine NIH 3T3 cells, the DNA is generally maintained as extrachromosomal elements and is therefore unstable in the absence of selection. The use of mutated DHFR with reduced affinity for methotrexate, while facilitating primary selection with methotrexate in cells with endogenous DHFR, is not particularly useful for subsequent amplification because of the very high concentrations of methotrexate required.

6.4. Potential Problems

In seeking to attain high-level expression in mammalian cells, an important problem that may be encountered is toxicity of the protein of interest to the expressing cells. This could be revealed by finding fewer than the anticipated number of colonies after methotrexate selection and no increase in expression at higher methotrexate levels due to uncoupling of the DHFR and target gene expression vectors. For example, the level of c-H-ras obtained by methotrexate selection was considerably higher than that of activated (valine-12) H-ras, but was nonetheless refractory to amplification at methotrexate concentrations higher than 3 μM [Ricketts and Levinson, 1988]. It is therefore helpful to record the number of colonies obtained and determine the level of expression of the target protein at each level of selection. An approach to circumvent the problem of toxicity is to use an inducible promoter for expression of the gene of interest. A recently developed mammalian cell expression system, using an ecdysone-inducible promoter to drive the gene of interest (Invitrogen, Cat. # K1000-01), could prove very useful to limit expression of a harmful gene product until after addition of the inducing agent to the culture medium.

REFERENCES

Alt FW, Kellems RE, Bertino JR, Schimke RT (1978): Selective multiplication of dihydrofolate reductase genes in methotrexate-resistant variants of cultured murine cells. J Biol Chem 253: 1357–1370.

Ausubel FM, Brent R, Kingston RE, Moore DD, Seidman JG, Smith JA, Struhl K (eds) (1996): "Current Protocols in Molecular Biology." New York: John Wiley & Sons.

Christman JK, Gerber M, Price PM, Flordellis C, Edelman J, Acs G (1982): Amplification of expression of hepatitis B surface antigen in 3T3 cells cotransfected with a dominant-acting gene and cloned viral DNA. Proc Natl Acad Sci USA 79: 1815–1819.

Cohick WS, Clemmons DR (1994): Enhanced expression of dihydrofolate reductase by bovine kidney epithelial cells results in altered cell morphology, IGF-I responsiveness, and IGF binding protein-3 expression. J Cell Physiol 161: 178–186.

Freshney RI (1994): "Culture of Animal Cells: a Manual of Basic Technique." New York: John Wiley & Sons.

Hudziak RM, Schlessinger J, Ullrich A (1987): Increased expression of the putative growth factor receptor p185[HER2] causes transformation and tumorigenesis in NIH 3T3 cells. Proc Natl Acad Sci USA 84: 7159–7163.

Hussain A, Lewis D, Sumbilla C, Lai L, Melera PW, Inesi G (1992): Coupled expression of Ca^{2+} transport ATPase and a dihydrofolate reductase selectable marker in a mammalian cell system. Arch Biochem Biophys 296: 539–546.

Isola LM, Zhou S-L, Kiang C-L, Stump DD, Bradbury MW, Berk PD (1995): 3T3 fibroblasts transfected with a cDNA for mitochondrial aspartate aminotransferase express plasma membrane fatty acid-binding protein and saturable fatty acid uptake. Proc Natl Acad Sci USA 92: 9866–9870.

Israel DI, Nove J, Kerns KM, Moutsatsos IK, Kaufman RJ (1992): Expression and characterization of bone morphogenetic protein-2 in hamster ovary cells. Growth Factors 7: 139–150.

Jou Y-S, Matesic D, Dupont E, Lu S-C, Rupp HL, Madhukar BV, Oh SY, Trosko JE, Chang C-C (1993): Restoration of gap-junctional intercellular communication in a communication-deficient rat liver cell mutant by transfection with connexin 43 cDNA. Mol Carcinog 8: 234–244.

Kaufman RJ, Sharp PA (1982): Amplification and expression of sequences cotransfected with a modular dihydrofolate reductase complementary cDNA gene. J Mol Biol 159: 601–621.

Kaufman RA, Sharp PA, Latt SA (1983): Evolution of chromosomal regions containing transfected and amplified dihydrofolate reductase sequences. Mol Cell Biol 3: 699–711.

Kaufman RJ (1990a): Selection and coamplification of heterologous genes in mammalian cells. Methods Enzymol 185: 537–566.

Kaufman RJ (1990b): Vectors used for expression in mammalian cells. Methods Enzymol 185: 487–511.

Kaufman RJ, Davies MV, Wasley LC, Michnick D (1991): Improved vectors for stable expression of foreign genes in mammalian cells by use of the untranslated leader sequence from EMC virus. Nucleic Acids Res 19: 4485–4490.

McCartney JE, Tai MS, Hudziak RM, Adams GP, Weiner LM, Jin D, Stafford WS, Liu S, Bookman MA, Laminet AA, Fand F, Houston LL, Oppermann H, Huston JS (1995): Engineering disulfide-linked single-chain Fv dimers [(sFv')$_2$] with improved solution and targeting properties: anti-digoxin 26-10 (sFv')$_2$ and anti-c-erbB-2 741F8 (sFv')$_2$ made by protein folding and bonded through C-terminal cysteinyl peptides. Protein Eng 8: 301–314.

Murray MJ, Kaufman RJ, Latt SA, Weinberg RA (1983): Construction and use of a dominant, selectable marker: a Harvey sarcoma virus-dihydrofolate reductase chimera. Mol Cell Biol 3: 32–43.

Ricketts MH, Levinson AD (1988): High level expression of c-H-ras1 fails to fully transform Rat-1 cells. Mol Cell Biol 8: 1460–1468.

Ricketts MH, Vandenplas S, van der Merwe T, Bellstedt D, North HM (1991): Analysis of co-transfected expression vectors amplified by methotrexate selection in Rat-1 cells. Eur J Cell Biol 56: 43–48.

Ricketts MH, Chiao E, Hu MC, Chung BC (1992): Amplification of P450c21 expression in cultured mammalian cells. Biochem Biophys Res Commun 186: 426–431.

Ringold G, Dieckmann B, Lee F (1982): Co-expression and amplification of dihydrofolate reductase cDNA and the Escherichia coli XGPRT gene in Chinese hamster ovary cells. J Mol Appl Genet 1: 165–175.

Schimke RT (1988): Gene amplification in cultured cells. J Biol Chem 263: 5989–5992.

Simonsen CC, Levinson AD (1983): Isolation and expression of an altered mouse dihydrofolate reductase cDNA. Proc Natl Acad Sci USA 80: 2495–2499.

Southern PJ, Berg P (1982): Transformation of mammalian cells to antibiotic resistance with a bacterial gene under control of the SV40 early promoter. J Mol Appl Genet 1: 327–341.

Stromqvist M, Andersson JO, Bostrom S, Deinum J, Ehnebom J, Enquist K, Johansson T, Hansson L (1994): Separation of active and inactive forms of recombinant human plasminogen activator inhibitor type 1 (PAI-1) expressed in Chinese hamster ovary cells: comparison with native human PAI-1. Protein Expr Purif 5: 309–316.

Tropp WC (1981): Normal rat cell lines deficient in nuclear thymidine kinase. Virology 113: 408–411.

Urlaub G, Chasin LA (1980): Isolation of Chinese hamster cell mutants deficient in dihydrofolate reductase activity. Proc Natl Acad Sci USA 77: 4216–4220.

Wigler M, Silverstein S, Lee L-S, Pellicer A, Cheng Y-C, Axel R (1977): Transfer of purified Herpes virus thymidine kinase gene to cultured mouse cells. Cell 11: 223–232.

Wigler M, Perucho M, Kurtz D, Dana S, Pellicer A, Axel R, Silverstein S (1980): Transformation of mammalian cells with an amplifiable dominant-acting gene. Proc Natl Acad Sci USA 77: 3567–3570.

APPENDIX: MATERIALS AND SUPPLIERS

Materials	Suppliers
Adenosine	Sigma
ADT medium: α-medium without nucleosides (#32561)	Life Technologies
Amethopterin	Sigma
Cloning cylinders	Bellco Glass
Deoxyguanosine (thymidine)	Sigma
Expression cloning vectors	Clontech, Invitrogen, Stratagene
F12:DMEM (1:1, Ham's F-12 and Dulbecco's modification of MEM)	Life Technologies
F12:DMEM without glycine, hypoxanthine and thymidine (#12500)	Life Technologies
Fetal bovine serum	Life Technologies
FBS, heat inactivated (# 10440-014)	Life Technologies
Filters, Millipore	Millipore
Geneticin (G418)	Life Technologies, Invitrogen
Methotrexate (+)amethopterin	Sigma
PBSA	Life Technologies
Penicillin/Streptomycin	Life Technologies
Petri dishes	Becton Dickinson

APPENDIX (*Continued*)

Plasmid isolation kit (#12143)	Qiagen
Plastic tubes, clear, 12mm × 75mm (#2058)	Becton Dickinson
pSV2Neo vector	Clontech Laboratories Inc.
Rat-2 cells	ATCC
Spiked polypropylene tube holder (#14-781-15)	Fisher
Trypsin/EDTA	Life Technologies

List of Suppliers

American Bioanalytical
10 Huron Drive, Natick, MA 01760
(800) 443-0600; fax (508) 655-2754

American Type Culture Collection
12301 Parklawn Dr., Rockville, MD 20852-1776
(800) 638-6597; fax (301) 231-5826

Amersham International
Amersham Place, Little Chalfont, Amersham, Buckinghamshre HP7 9NA, UK
01494 544000; fax 01494 542929

Amersham Life Science
2636 South Clearbrook Drive, Arlington Heights, Illinois 60005
(708) 593-6300; fax (708)-437-1640

AMS Biotechnology
12 Thorney Leys Park, Witney, Oxon OX8 7GE, England
01993 706500; fax 01993 706006

Apple Scientific Inc
8378 Mayfield Road, Chesterland, Ohio 44026
(216) 729-3056; fax (216) 729-0928

Beckman Instruments Inc
2500 Harbor Blvd., Fullerton, CA 92634
(800) 854-8067, (714) 871-4848; fax (714) 773-8283

Becton Dickinson Labware
2 Bridgewater Lane, Lincoln Park, New Jersey 07035; Two Oak Park, Bedford, MA 01730
(800) 235-5953, (800) 343-2035; fax (800) 847-2220

Bellco Glass, Inc
P.O. Box B340, Edrudo Rd, Vineland, NJ 08360
(800) 257-7043; fax 609-691-3247

BioFluids, Inc
1146 Taft Street, Rockville, MD 20850
(301) 424-3619; fax (301) 424-3619

Bio-Rad Laboratories
2000 Alfred Nobel Drive, Hercules, CA 95457
(800) 424-6723; fax (800) 879-2289; (510) 741-5800

Boehringer-Mannheim Biochemicals
9115 Hague Road, P.O. Box 50414 , Indianapolis, IN 46250-0414
(800) 262-1640; fax (800) 428-2883

Boehringer Mannheim
31 Victoria Avenue, P.O. Box 955, Castle Hill, NSW 2145
(02) 9899-7999; fax (02) 9634-2949

DNA Transfer to Cultured Cells, Edited by Katya Ravid and R. Ian Freshney.
ISBN 0-471-16572-7 © 1998 Wiley-Liss, Inc.

Charles River Breeding Laboratories
215 Ballardvale Street, Wilmington, MA 01887
(800) 522-7287; fax (508) 657-5012

Chiron Viagene
11055 Roselle Street, San Diego, CA 92121-1290
(619) 452-1288

Clontech Laboratories
1020 East Meadow Circle, Palo Alto, CA 94303-4230
(415) 424-8222; fax (415) 424-1064

Corning Costar Corporation
One Alewife Center, Cambridge, MA 02140
(800) 492-1110; fax (800) 358-5287

Corning, Inc.
Houghton Park E Bldg, Museum Way, Corning, New York 14831
(800) 222-7740; fax (617) 868-2076

Difco Laboratories
P.O. Box 331058, Detroit, MI 48232-7058
(800) 521-0851, (313) 462-8500; fax (313) 462-8500

DuPont NEN
549 Albany Street, Boston, MA 02118
(800) 551-2121; fax (800) 666-6527

East Coast Biologics, Inc
P.O. Box 489 North Berwick, ME 03906-0489
(207) 384-4431; fax (207) 384-4437

Eastman Kodak Company, Scientific Imaging Systems Division
343 State St., Rochester, NY 14652-4115
(800) 225-5352, (203) 786-5600; fax (800) 879-4979

Falcon
(see Becton Dickinson)

Fisher Scientific
711 Forbes Ave., Pittsburgh, PA 15219-4785
(800) 766-700; fax (800) 926-1166

Fluka Chemical Corporation
980 South 2nd Street, Ronkonkoma, NY 11779-7238
(800) FLUKAUS, (516) 467-0980; fax (516) 467-0663

Gelman Sciences, Inc
600 S Wagner Road, Ann Arbor, MI 48103
(800) 521-1520, (313) 665-0651; fax (313) 913 6440

Gen-Probe Inc
9880 Campus Point Drive, San Diego, CA 92121
(800) 523-5001, (619) 546-8000; fax (619) 452-5848

Harlan Bioproducts, Inc
P.O. Box 29176 Indianapolis, IN 46229-0176
(800) 924-362; fax (317) 894-1840

ICN Biomedicals Inc
1263 South Chillicote Road, Aurora, Ohio 44202
(800) 854-0530; fax (800) 334-6999

Intergen Company
2 Manhattan Ville Road, Purchase, New York 10577
(800) 431-4505; fax (914) 614-1429

International Biotechnologies, Inc (IBI)
New Haven, CT06535 (see also Eastman Kodak Co)
(800) 243-2555; fax (203) 786-5600

Invitrogen (see R&D Systems for Europe)
3985 B. Sorrento Valley Boulevard, San Diego, CA 92121
(800) 955-6288; fax (619) 597-6201

JRH Biosciences Inc (see AMS Biotech. for UK)
13804 W. 107th Street, P.O. Box 14848, Lenexa, KS 66215
(800) 255-6032; fax (913) 469-5584

Life Technologies
P.O. Box 68, Grand Island, NY 14072-0068
(800)828-6680; fax (800)331-2286

Life Technologies, Inc (GIBCO-BRL)
8400 Helgerman Ct., P.O. Box 6009, Gaithersburg, MD 20884
(800) 828-6866, (301) 840-8000; fax (800) 331-2286

Life Technologies Ltd, (GIBCO BRL) UK
3 Fountain Drive, Inchinnan Business Park, Paisley PA4 9RF
0141 814 6100; fax 0141 814 6317

Nalge Company
P.O. Box 20365, Rochester, NY 14602-0365
(716) 264-3985; fax (800) 625-4363

Nalge-Nunc International
2000 North Aurora Road, Naperville, IL 60563-1796
(800)NALGE-CF, (800) 625-4327; fax (800) 625-4363

New England Biolabs
32 Tozer Road, Beverly, MA 01915-5599
(800)NEB-LABS, (508) 927-5054; fax (508) 921-1350

Nichols Institute Diagnostics
33608 Ortega Highway, San Juan Capistrano, CA 92690-6130
(800) NICHOLS, (714) 728-4000; fax (714) 728-4970

Nunc
Postbox 280, Kamstrup, DK 4000 Roskilde, Denmark
45 46 35 9065; fax 45 46 35 0105

Pharmacia Biotech, Inc
800 Centennial Ave., P.O. Box 1327, Piscataway, NJ 08855-1327
(800) 526-3593, (908) 457-8000; fax (800) FAX-3593

Pharmingen USA (AMS Biotech UK)
11555 Sorrento Valley Road, Suite E, San Diego, CA 92121
619 792 5730; fax 619 792 5238

Promega
2800 Woods Hollow Road, Madison, WI 53711-5399
(800) 356-9526, (608) 274-4330; fax (800) 356-1970; (608) 277-2516

Qiagen Inc
9600 De Soto Avenue, Chatsworth, CA 91311
(800) 426-8157, (818) 718-9870; fax (800) 718-2056

Qiagen Ltd, UK
Unit 1, Tillingbourne Court, Dorking Busines Park, Dorking Surrey RH4 1HJ, UK
01306 740444; fax 01306 875885

R & D Systems Europe
4–10 The Quadrant, Barton Lane, Abingdon OX14 3YX
01235 531074; fax 01235 533420

Research Products Inc
410 N Business Center Drive, Mount Prospect, Illinois 60056
(708) 635-7330; fax (708) 635-1177

Sigma Chemical Company
P.O.Box 14508, St Louis, MO 63178
(800) 325-3010; fax (800) 325-5052

Stratagene
11011 North Torrey Pines Road, La Jolla, CA 92037
(800) 424-5444; fax (619) 535-0045

Syntex Chemicals Inc
2075 N. 55th St., Boulder, CO 80301
(303) 442-1926; fax (303) 938-6413

Upstate Biotechnology, Inc
89 Saranac Ave. Lake Placid, NY 12946
(800) 233-3991; fax (617) 890-7738

VWR Scientific
1310 Goshen Pkwy., W. Chester, PA 19380
(800) 932-5000, (610) 936-1700; fax (610) 936-1761

Worthington Biochemical Corporation
Halls Mill Road, Freehold, NJ 07728
(800) 445-9603; fax (800) 368-3108

Index

protein expression analysis, 275–276
reagents and media, 267–271
transfection and selection of cells, 271–275
Dimethylsulfoxide, *See* Dimethylsulfoxide
Dimethylsulfoxide (DMSO)
in calcium phosphate transfection, 116, 143
in DEAE-dextran transfection, 180
DMEM
adipose cell electroporation, 95
DEAE-dextran in, 181, 185–187
gpt selection media in, 7
high glucose, 6–7
lipofection, 195–196, 203
zebrafish culture, 78
DMEM/10FB, for calcium phosphate transfection, 160
DNA, *See* Plasmid DNA; Transfection
Double-antibody assay, for cell surface epitope-tagged GLUT4, 101–103
Dropout mix powder and medium, 226–227
Drug resistance, *See* Selection
Dulbecco's modified Eagle's medium, *See* DMEM entries

Early to late (ETL) promoter, 31–32
ECV, *See* Extracellular virus particles
Electrophoresis, *See* Gel electrophoresis
Electroporation
comparison with calcium phosphate transfection, 158–159
into cultured cell lines
background and theory, 56–57
comparison with other methods, 58–59
critical parameters for efficiency, 57–58
megakaryocytic cell lines, 62–64
reporter gene expression detection, 64–65
suppliers, 66–67
vectors, reagents, and supplies, 59–62
of rat adipose cells
adipose cell preparation, 96–97
applications, 94
culture, 99
cuvette preparation and use, 98–99
double-antibody assay for GLUT4, 101–103
D[U-^{14}C]glucose uptake assay, 99–101
insulin signal transduction, 103–107
lipid content determination, 103
materials and suppliers, 108–109
plasmid DNA preparation, 97–98
reagents, 95–96

Embryonic stem cells (ES cells)
comparison with zebrafish cultures, 73–74, 88–89
retrovirus vectors for, 2
Embryos, *See* Zebrafish embryo cell cultures
Enhancers, *See* Promoters
env gene, 4–5, 7, 76
Envelope proteins
fusogenic, 8, 12
for retroviral vector production, 6
Enzymes
baculovirus studies, 45–50
heparan sulfate D-glucosaminyl N-deacetylase/N-sulfotransferase type I and II, 190
restriction, 223–225
Epididymal fat pad dissection, 96
Epitope-tagged GLUT4, 101–103
Epstein-Barr virus sequences, 10
ES cells, *See* Embryonic stem cells
Ethidium bromide, 227
ETL promoter, *See* Early to late promoter
Evans-Kaufman medium, 228
Expression vectors, *See* Vectors
Extracellular matrix protein synthesis, zebrafish cultures, 72–73
Extracellular virus particles (ECV), baculovirus, 29–30

FBS, *See* Fetal bovine serum (FBS)
FCS, 143–144
Feed layers, zebrafish culture, 81
Fetal bovine serum, for virus production in ψ2 cells, 23
Fetal bovine serum (FBS)
for calcium phosphate transfection, 269
for DEAE-dextran transfection, 185–187
heat inactivated, for electroporation, 61
FISH, *See* Fluorescence in situ hybridization
Fluorescein isothiocyanate (FITC), 173–175, 207–209, 251–252
Fluorescence, *See* Immunofluorescent staining
Fluorescence in situ hybridization (FISH), 224, 249–252
Freezing Sf cells, 34
Functional assays, 259–260

G418, *See* Geneticin
G-11 staining, 248–249
gag gene, 4–5, 7
β-Gal assay kit, 139
β-Gal internal control plasmids, 127–128